10 STRATEGIES FOR SUCCESSFUL PROJECTS

Elevate your management to the next level

GERMAN RUMINOT

10 strategies for successful projects
Elevate your management to the next level - German Ruminot

ISBN: 9798335027939
First edition, august 2024

For anyone seeking to enhance their project management skills!

TABLE OF CONTENTS

PROLOGUE

In this book, the author presents **ten strategies** in a straightforward and practical manner to enhance the management of technological projects. The book is supplemented with examples and case studies, enabling readers to readily and effectively apply these strategies to their own projects.

This book examines and elaborates on key concepts, including the project's purpose as a central focus for teams and stakeholders, and the critical role of communication in remote and hybrid environments. It also explores how these factors contribute to the analysis of project complexity.

Chapter one analyzes the different *types of projects* that coexist today in Project Management.

Chapter Two explores the concept of Areas, which serves as an essential foundation for defining the environment and complexity of your project.

Chapter Three develops the idea of how to structure Autonomous Teams to increase the value generated to the project.

Chapter Four addresses customer management, effectively defining strategies that will help you develop this skill.

Chapter Five deals with the strategy for *communicating effectively* to all areas of the project.

Chapter Six details the strategy for *effective requirements* gathering, which will allow you to define the correct scope for your project.

Chapter Seven discusses a strategy that proposes a change of vision in the project to focus on results and customer-perceived value.

The following chapters address strategies for dealing with risks, quality and expenses, with the aim of maintain these three key areas of your project under control and properly managed.

Finally, we present a contingency plan with sample templates, so that you can apply each of the strategies quickly and agilely to your project.

INTRODUCTION

Following the coronavirus pandemic, Project Management had to implement significant changes to ensure its effective application globally. It has become a strategic asset for companies, addressing the evolving demand, health restrictions, and intense competition across various markets.

This new reality has compelled organizations to rapidly adapt by incorporating innovations and continuous improvements into their products and services to capture emerging trends in purchasing and usage. These new initiatives are implemented through projects, which must be managed effectively to achieve the set objectives

For this reason, in this release we propose a series of *strategies, which will help you improve* the *Management of your Technological Projects,* understanding by strategy the set of actions that search to achieve a specific objective. In our context, the objective sought is the success of your project so that the client achieves the expected value.

Throughout this book, we will provide a practical guide to incorporate these *strategies*, achieving excellent results in a very short time, taking your *Project Management* to the next level of excellence.

Join us to review each of these strategies that we present below.

DISCOVERING THE TYPES OF PROJECTS

We will begin our exploration with a brief introduction to the various types of projects you may encounter in Project Management.

It is important to note that while there are multiple ways to categorize projects, for our analysis, we will focus on these three categories:

- o   Adaptive Projects (Agile)
- o   Predictive Projects (Waterfall)
- o   Hybrid Projects

Note:

It is important to note that in the current business environment, an increasing number of companies are shifting from Predictive Projects to Adaptive and Hybrid Projects, where the project evolves in response to changing conditions and client needs.

Adaptive Projects (Agile)

These types of projects are characterized by being developed in environments with very changing demands.

The premise of this type of project is to point out the following:

"With the given budget, deliver the most results in the shortest possible time."

Therefore, the scope will increase as the results are delivered over time.

These types of projects are used when the following characteristics are identified in the context:

- **Frequent changes** in the project.
- **Changing** business **demands**.
- **Autonomous teams** with a high degree of self-control.
- **High participation and collaboration of the client and stakeholders** in the definition of the results to be released, throughout the development of the project.
- Practice is prioritized over the detailed monitoring of a methodology.
- Defined time windows are used to release deliverables and results to the client.
- Changes are welcome and their management is simplified, given the context where the project is developed.

In general, the work cycle in this type of project considers:

○ Generate deliverables that add value to the customer through defined time windows, through a set of successive iterations. After each iteration, the customer will receive the results. For each iteration, new deliverables are defined to be performed.

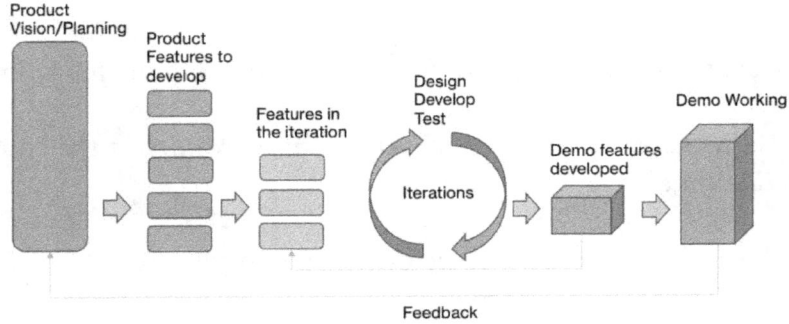

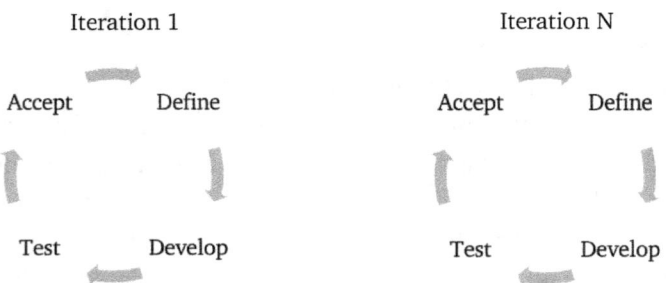

Figure 1 Agile Project

Predictive Projects (Waterfall)

These projects are characterized by being developed in environments with defined demands, which are specified at the beginning of the project, and the value of the results, the client perceives them at the end of this.

The premise of this type of project is defined with the following:

"With the budget, time and scope defined at the beginning of the project, deliver at the end of the project, the required product that meets the agreed specifications."

Therefore, the scope is defined at the beginning of the project and the benefit of the product is perceived at the end of it.

An example of this type of project is for example the construction of a bridge, building or infrastructure, where the final deliverable can undergo very few major changes.

Some characteristics of this type of projects are:

- o **Scope, time, and budget defined** at the beginning of the project.
- o **Strict control of changes,** so that they do not affect the scope of the project.
- o **Rigorous monitoring of the methodology** for the development of the project.
- o **Team organized hierarchically and with well-defined roles.**

o The client takes a leading role at the beginning of the project, for the definition of requirements.

In general, the work cycle in this type of project is:

o Define requirements at the beginning of the project, analyze and develop them, to later test and deliver the product to the client in the defined time and budget. Any change must be defined outside the initial scope, considering a specific time and budget for it.

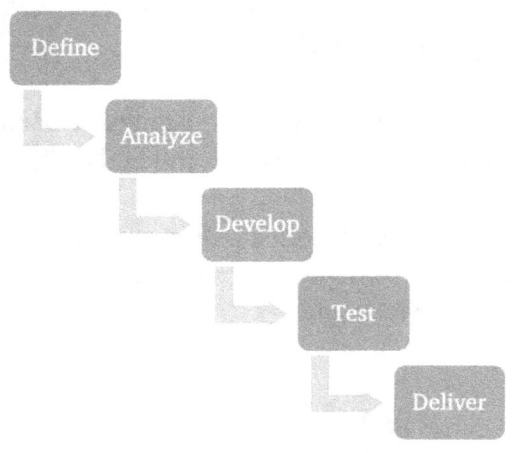

Figure 2 Waterfall Project

Hybrid Projects

This type of project is the result of mixing an *Adaptive project* and *Predictive project*.

The objective is to deliver value to the client before the end of the project, through partial deliveries of the product, maintaining the definition of requirements and scope at the beginning of the project.

This type of project can be a good transition to move from a *Predictive* project to an *Adaptive one gradually, when organizations, teams and clients are not fully adapted to projects of the* Adaptive type.

The work cycle, in this type of projects considers the following steps:

- o Define requirements and scope at the beginning of the project, **analyze and prioritize the results in iterations**, then develop them, and later test and deliver the products to the client in each of the iterations. As in the *Predictive* project, any change must be defined outside the initial scope, considering a specific time and budget to incorporate it into any of the subsequent iterations.

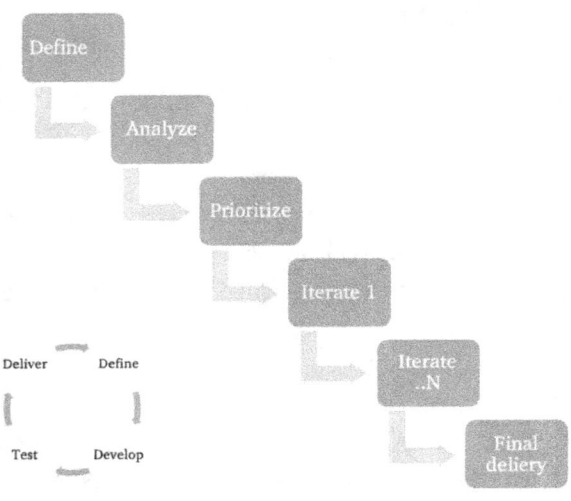

Figure 3 Hybrid Project

Having reviewed the different types of projects, we will now explore the strategies you can apply to enhance the management of your projects.

STRATEGY 1: *DEFINE YOUR SCOPES AND COMPLEXITY*

"First analyze your environment, then plan your path."

Nowadays, projects are increasingly complex and dynamic to manage, for this reason, through this strategy we define as a first step the measure of complexity that your project will have, prior to the start of its development.

For this process, we will make a complete analysis of the environment and the different actors that will participate in the project. Starting by defining some essential concepts that will allow you to describe and categorize your project.

Areas

We will understand by areas, the various dimensions that will influence a project throughout its development. These influences can be decisive to achieve the success of your project.

It is crucial to conduct an analysis of the project's areas before commencement to establish an initial complexity context. This will facilitate effective management and control in subsequent stages.

Here are the influences areas you should identify in your project:

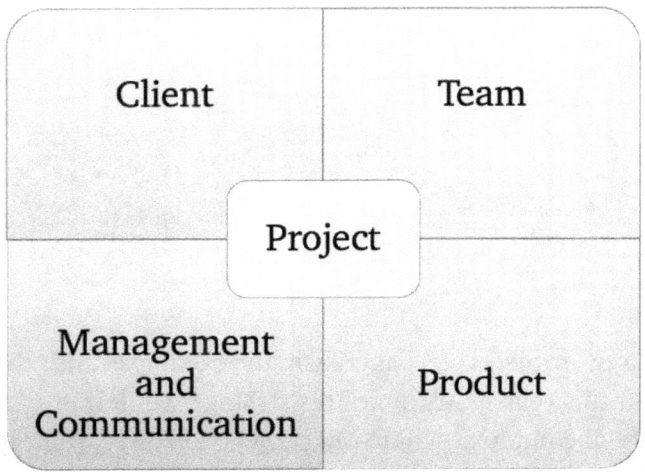

Figure 4 Influences Areas of a Project

Client Scope

This area defines all relevant actors interacting with the project. Understanding by client, such as sponsors, users, and other actors, who can directly or indirectly influence the development of the project.

The steps you must follow to define the scope of the client consider:

    a. Identify your clients and their grade of decision and influence on the project.

    b. Identify your structure and processes for decision making.

c. Identify *who* and *how* will make the decisions in your project, asking yourself, for example:

> o   Who can change or add a requirement?
> o   How is the process for changing a requirement in the client?
> o   Who can validate and accept a deliverable in the project?

d. Identify for each of the clients the following characteristics:

> o   Authority
> o   Role in the project
> o   Interests
> o   Expectations
> o   Knowledge and influence on the project

e. Identify if you have relocated customers, this means in different locations:

> o   Branch, Region, or Country
> o   Different buildings
> o   Interaction between each customer

Team Scope

Defining the team's scope provides a clear understanding of its composition, organizational structure, and interaction with the client. For which you must perform the following steps:

a. List your team members and the role they will take in the project.
b. Defines the structure and experience of team members.

   c. Detail how they will interact to provide some response
       to your project and the client.
   d. Detail if you have a team with external supplier and
       their service levels agreement that you will demand.
   e. Detail the experience needed and what member in your
       team possess for each role in the project.
   f. Finally, it is important to define where the team will be
       distributed in person, hybrid or remotely.

Management and Communication

Establishes the management and communication processes for both the client and team scopes of the project. This involves listing all procedures and communications required throughout the project, both internally and with the client, specifying their frequency and identifying the participants for each.

The coronavirus pandemic marked a significant turning point in management and communication practices, necessitating a shift from entirely in-person interactions to remote or hybrid models. This transition had to be achieved while still meeting expected results within a more constrained timeframe.

Product Scope

This section must detail the processes for defining, controlling, validating, and approving product deliverables. Additionally, it should specify the type of product to be deployed, whether it is an existing product in the market, an enhancement to an existing product, or an innovation new to the market.

Project Complexity

Then, having completed each of the areas mentioned above, you are ready to analyze and assign an initial level of complexity to your project.

To analyze the complexity of the project, a set of characteristics have been defined for each area, which will allow you to measure its impact on the project.

The review the project complexity is dynamic during each stage of the project. After each review it will be necessary to generate actions to correct the course of the project.

Presented below are the analysis tables that will be used to illustrate the concept of project complexity:

Client Scope

Characteristics	Complexity
o The *Client* is arranged in the same location. o The internal structure of the project (well-defined roles) belongs to the same area of the company. o *Users and Sponsors* are experts in the area of project development.	1 low
o The *Client* is arranged in the same location. o The internal structure for decision-making involves a single area of the company. o *Users and Sponsors* are experts in the area of project development.	2 medium

- o The *Client and sponsor* are distributed in different locations.

- o The internal structure of the *Client* for decision making involves several areas of the company (commercial, finance, etc.).

  3 high

- o *Users and Sponsors* are beginners in the business sector and project development concepts.

Team Scope

Characteristics	Complexity
o The entire *team* is in the same location and in person. o *Expert team* in technical and management issues. o *Team* without external suppliers.	1 low
o The entire *team* is in the same location and works in a hybrid modus. o *Team* mixes of inexperienced staff with experts, in technical and management issues. o *Team* without external suppliers.	2 medium
o *Team* distributed in different locations and works 100% remotely. o *Team* mixes of inexperienced staff with experts, in technical and management issues. o *Team* without external suppliers.	3 high

- o *Team* distributed in different locations and 100% remote work.   4 very high
- o *Inexperienced team* in technical and management issues.
- o *Team* with external suppliers.

Management and Communication

Characteristics	Complexity
o Communication occurs at a single location for both the entire team and the client. o There is no *communication* with external suppliers.	1 low
o The *Communication* takes place in the same location for the team and another location for the client and stakeholders. o *Communication* with external suppliers.	2 medium
o *Remote communication* in different locations for the team and the client. o *Communication* with external suppliers. o *Mostly remote* communications.	3 high

Product Scope

Characteristics	Complexity
o The *Product* is developed in a mature and stable technology. o The *Product* will not integrate with other products or systems. o The *Product* does not support a critical business process. o The Product already exists in the market or company.	1 low
o The *Product* will be developed in a mature and stable technology. o The *Product* will integrate with other systems. o The *Product* does not support a critical business process. o The product to be released is an improvement to one already existing in the market or the company.	2 medium
o The *Product* is developed in a mature and stable technology. o *The Product* integrates with other systems. o The *Product* will support a critical business process. o The product is an innovation for the market.	3 high

Determine Complexity

Upon completing the complexity sheets for each scope, we will sum these values linearly to determine the overall complexity value of the project, as demonstrated below.

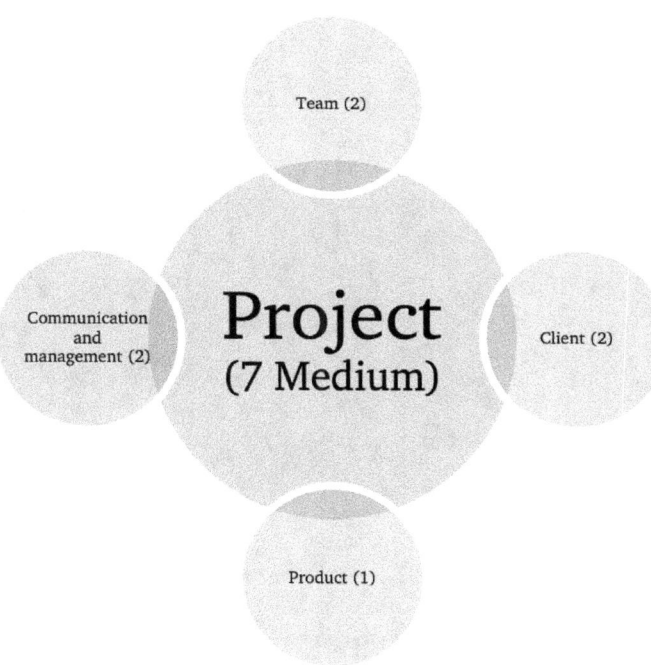

Figure 5 Project Complexity

Note

If in you case you consider that any of the areas has a greater relative weight with respect to the rest, you can apply a different factor in the sum of the values to calculate the complexity of your project.

Complexity ranges

Next, we define the ranges of complexity in which the project can vary, for its subsequent classification:

- 4 and 6   *Low*
- 7 and 8   *Medium*
- 9 and up   *High*

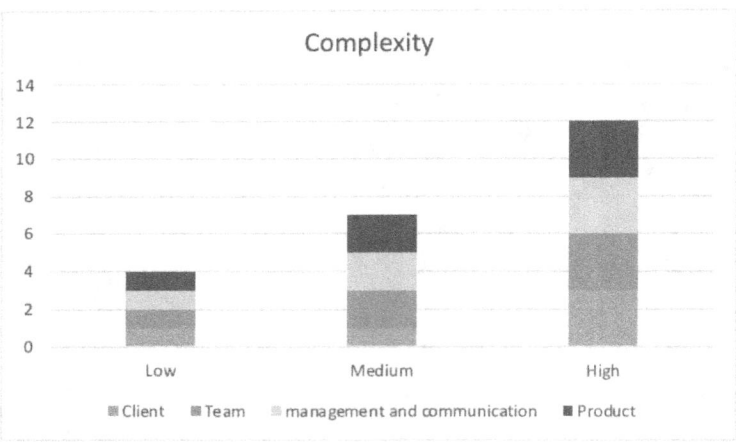

Figure 6 Complexity by Areas

Determining the complexity of your project is crucial for focusing management strategies on areas with high complexity scores. These areas, if not managed effectively, can become significant management challenges during the project's development.

Next, we will analyze a case study to determine the scope and complexity of a project.

Case Study

Below is an example illustrating the process of defining the scopes and assessing the complexity of a project.

RET INC

The company *OGP INC.* has commissioned the company *RET INC.* the development of a product for the Commercial Department.

This product will have the function of predicting the conversion effectiveness of prospects into customers, through the exhaustive analysis of the characteristics of each of the prospects to become customers of the company.

The cycle of conversion of a prospect into a customer is estimated at 3 months since it enters the database of the Commercial Department.

This new product will be called *PROSPECT*.

OGP INC. has determined that:

- o The project will be sponsored by the Chief Financial Officer (Hans Blue) from Chicago.
- o Users of the system (Artur Klo, Bennet Hiu) members of the Commercial Department will be available for this project from Atlanta.
- o The internal project leader (Alexander Fle) will be from the Projects Department, available from Chicago.
- o The product must be integrated with the company's current ERP, specifically with the financial module.

    o   The product must operate on all mobile phones of the Commercial Department staff, both inside and outside the company.

    o   This product will be a critical system for the management of new customers of the Commercial Department.

Conversely, RET INC has determined the following:

— It will have a senior project leader (Greck Gast) in Chicago.
— A senior, technical and management team (Hortar Ju, Hans Ruls, Alex Bjo) in Atlanta.
— Initially it is proposed, not to have external suppliers to address this project.

Client Scope

We begin the analysis of the client's scope, defining the client's structure to address the project, which will give us a clear vision of how to manage the client in each situation of the project.

Name	Artur Klo	Alexander Fle	Bennet Hiu	Hans Blue
Role / Structure	Key User / Commercial Department	Project Leader / Department	User / Commercial Department	Sponsor/ Finance Department
Actions and Decisions	Define Requirements	Approves changes and deliverables, Manages the internal team	Test and validate Project Deliverables	Manage invoice and contract approval
Location	Atlanta	Chicago	Atlanta	Chicago

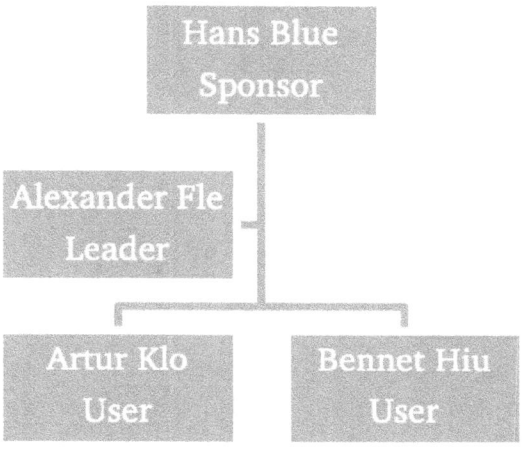

Figure 7 Client Scope

Team Scope

To define the team's scope, we will analyze several factors, including the roles, structure, and responsibilities of each member within the project.

Name	Hortar Ju	Hans Ruls and Alex Bjo	Greck Gast
Role/ Structure	Analyst- Consultant / depends on the Team Leader	Developers/ report to the Team Leader	Team Leader
		They code and test.	Manage and control tasks

Actions and Decisions	Analyzes and documents requirements with the client. Test the functionals released by the developers. Hold meetings with clients.	Holds meetings with Analysts and Consultants	and deliverables. Hold meetings with clients
Location	Atlanta	Atlanta	Chicago

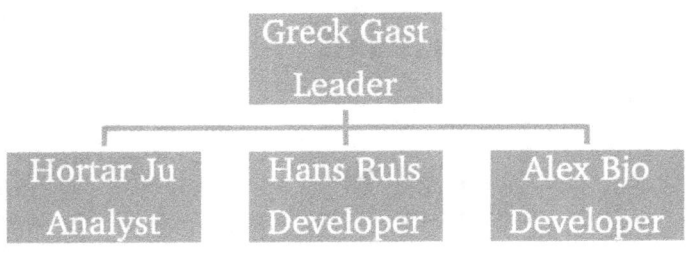

Figure 8 Team Scope

Management and Communication

In this area we will define all the meetings, communications, participants, and periodicity for the team and client during the project.

Meetings

Type of Meeting	Detail	Periodicity	Participants
Kickoff	Initial meeting of the project or iteration.	One (1) week after signing the contract, the kick-off meeting is defined or one (1) week after	All Team and Client

		the previous iteration ends.	
Weekly management meetings	Meetings to review statuses. of deliverables.	Every Wednesday (1 hour)	Team & Client
Internal control meetings	Task control: status, assignment, and end dates.	At the beginning and end of the week (30 min)	Team
Functionality and deliverables review meetings	The functionality and deliverables of the project are reviewed.	Biweekly (1 hour)	Team & Client
Risk Meeting	Risk control: quantity, allocation, and status.	At the beginning of each week (15 to 30 min)	Team
Requirements meetings	Sessions to define, validate and approve requirements.	At the beginning of each iteration or start of a project (1-hour sessions)	Team & Client
Work meetings	Meetings are defined to discuss particular topics of requirements or deliverables.	By demand of the Team or Client (1 hour)	Team and/or Client
1 to 1 meeting	Meeting to talk with each member of the team about the performance.	Team	Team

Communications

Communication	Detail	Emitter	Receptor
Meeting minutes	Meeting to review discussed topics, agreements made, and outstanding issues for the day.	Team	Customer
Requirements Document	A document detailing the requirements for the iteration or cycle of the project.	Team	Customer
List of prioritized deliverables	list of deliverables Prioritized by business value.	Customer	Team
Project Risk List	Risk control: quantity, allocation, and status.	Team	Customer
List of tasks of the week	List of current tasks for the project.	Team	Team
List of completed deliverables	List of deliverables to be approved.	Team	Customer

Product Scope

Next, we will define the procedures associated with the scope of the product.

Product Management	Detail	Periodicity	Responsible
1	Definition of deliverables.	They are defined at the beginning of each iteration or start of the project.	Client & Team
2	List of deliverables prioritized by business value.	They are defined at the beginning of each iteration or start of the project.	Customer
3	Development of deliverables.	During the iteration of the list of deliverables.	Team
4	Control of Deliverables.	Biweekly	Team & Client
5	List of completed deliverables to be approved.	At the end of the iteration cycle.	Team
6	List of approved deliverables.	At the end of the iteration cycle.	Customer
7	Lesson learned about the process and deliverables.	At the end of the iteration cycle.	Team

Complexity

After defining the areas, we will proceed to assess the complexity of the project.

Ambit	Analysis	Complexity
Customer	o *Client* distributed in multiple locations. o *Expert client* in the commercial area. o It interacts with several departments of the company (projects and Commercial).	3 high
Team	o *Team* distributed in several locations. o *Technical and management expert* team. o No third-party providers.	2 medium
Management and Communication	o *Remote communication* in different locations.	3 high
Product	o *Product* supports critical process for the Commercial Department. o The *Product* integrates with external systems. o The *Product* is developed on technology known to the team.	3 high

Next, we present graphically, the impact of each area on the complexity of the project.

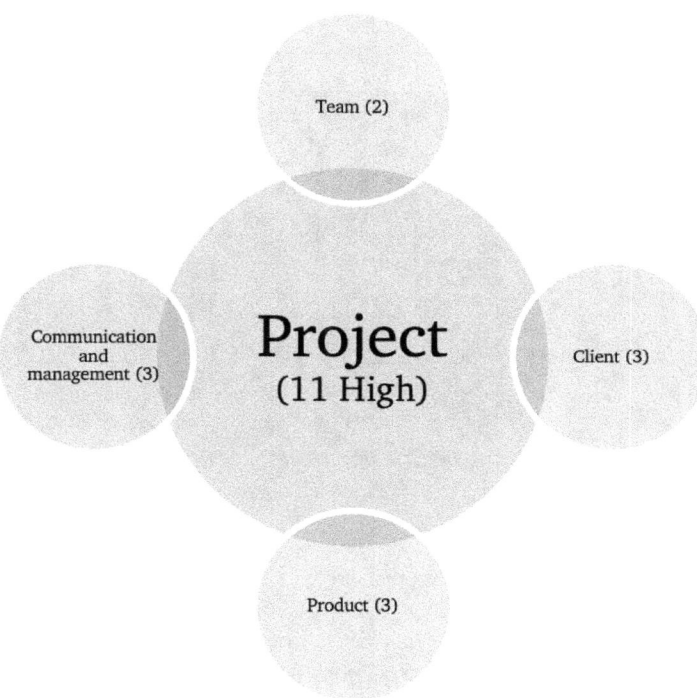

Figure 9 Project Complexity

In addition, we can visualize the detail of how each area impacts the overall complexity of the project.

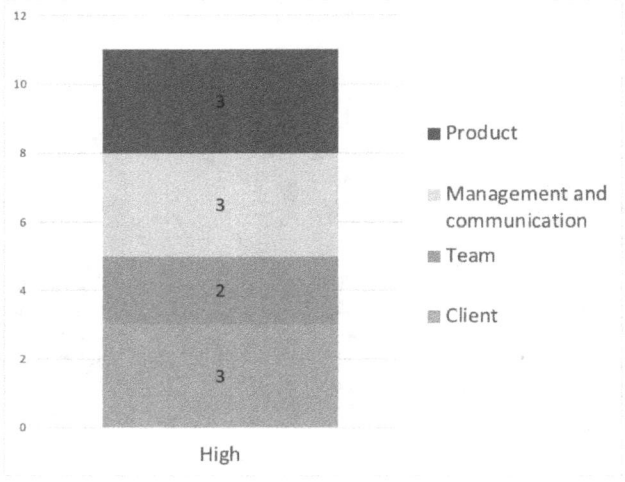

Figure 10 Detail of Project Complexity by Scope

Based on the previous analysis, it can be inferred that the project exhibits *high complexity* across nearly all areas. Therefore, it is crucial to implement strategies from the outset to address these complexities, in order to mitigate and control their impact throughout the project's development.

The following chapters will present specific strategies designed to address the various complexities within your project.

STRATEGY 2: CREATE AUTONOMOUS TEAMS

"Autonomous teams, more than an army without objectives."

In today's environment, characterized by dynamic needs and tight timelines for market entry, it is essential to have highly integrated and proactive teams. These teams must be capable of responding swiftly and effectively to the needs of customers and stakeholders.

Consequently, there is a need for high-performance **autonomous teams** that can simultaneously address the evolving requirements of a dynamic and demanding environment.

After the coronavirus pandemic, being able to count on high-performance **Autonomous Teams** took a preponderant role for companies, therefore, the creation of this type of teams is an essential process to achieve it.

Examining the implications of lacking an autonomous team reveals significant drawbacks, such as the presence of multiple hierarchical layers and distinctly separated roles. These factors result in cumbersome work structures and slow management processes, characterized by excessive supervision. Such conditions hinder the ability to respond swiftly and effectively to the evolving needs of clients and the market, ultimately failing to deliver the expected value to the business.

Characteristics of Autonomous Teams

There are fundamental characteristics, with which we can identify an *Autonomous Team*, among which are the following:

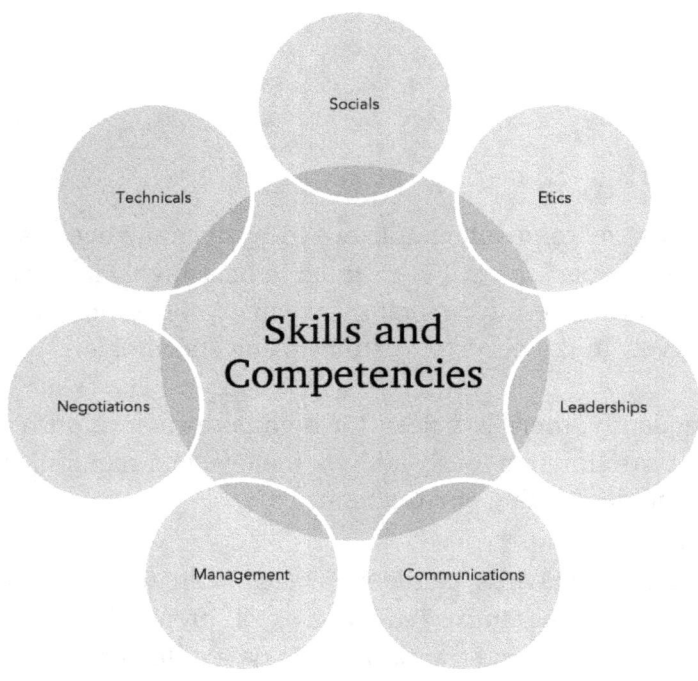

Figure 11 Skills and Competencies

The fundamental thing for the success of your project is *that the autonomous* team as a whole has these *skills and competencies, so that each of its members can contribute (their skills and competencies) to the team, thus enhancing each of the required actions.*

Next, we will examine the impact that the absence of any of these skills and competencies may have on the project.

Interpersonal skills

Lack of *social skills* can be perceived when:

- o Team members are isolated and will not interact with the rest, generating knowledge silos within the project.
- o The team cannot conduct a conversation or meeting autonomously with the customer.
- o Team members do not communicate or interact with each other, generating communication gaps, which can generate delays in deliverables and tasks,
- o Lost time and rework in the project.

Leadership Skills

The lack of *leadership skills in the* team manifests itself in the following situations:

- o The team can wander around without a definite direction.
- o There is no clarity of the objectives or purpose of the project, they will be like a herd without a clear direction.
- o Increase in misunderstandings and conflicts within the team and facing the client.
- o Internal and clear discoordination to the client.
- o Increases in development times in tasks and deliverables.
- o Rework.
- o Increases in project expenditures.

Negotiation and Communication Skills

The team must have *Negotiation and Communication skills* to address issues such as:

o   Listen and understand what is being communicated.
o   Draft and agree requirements or changes with the client.
o   Review and approve features of a product or deliverable.
o   Communicate clearly and effectively the messages you want to convey to the client or team.
o   Define scopes and negotiate deliverable dates with the client.
o   Interact correctly among all team members.

Management Skills

These types of competencies have an impact on the management of tasks and deliverables.

o   The team, not being able to manage and administer their own tasks, will not be clear about which are priorities and in what order they should be performed.

o   There could be discoordination in the team and facing the client.

o   There is no self-management of each of the team members to address their commitment and project tasks.

Technical Skills

This lack of technical skills will be evident in the quality of the deliverables, given that, from a product point of view, the team will not have the required knowledge and technical capabilities.

The impact of not having these characteristics can be reflected in:

- o Increases in the time required to perform tasks.
- o Rework.
- o Do not end up determined by lack of competences.
- o Increased project expense and risks.
- o Problems of acceptance of the deliverable or product by the client.
- o Non-compliance with deadlines and estimates made.

Ethical competencies of the team.

This is a crucial competence that every team member involved in a project must possess. It pertains to an ethical foundation grounded in principles of justice, the common good, and respect for the dignity of individuals practicing their profession.

Therefore, if ethical breaches are detected in the team, they must be eradicated immediately, first, for the well-being of the team, and second the normal development of the project.

The lack of ethics of any of the team members can generate great problems for the team and the project if they are not detected early.

The recommendation in this type of situation is to analyze the order of magnitude of the fault. If it is of a significant magnitude and puts the project at risk, the most appropriate action will be to change the member who has committed the fault, for the good of the team and the success of the project.

Advantages of Autonomous Team

The primary advantages of having autonomous teams are:

1. Increases team efficiency and productivity.

2. They optimize time, reducing expenses in non-productive hours and delays.

3. Increase in the value perceived by the client throughout the project, by releasing more deliverables with the highest expected value in the shortest time.

4. They improve the communication and synergies of the team, eliminating knowledge silos within the project.

5. Reduction of the risk associated with the scope of the project team.

6. They understand the purpose of the project and the value of its results to the client and the business.

Create your Autonomous Team.

The strategy of *Autonomous Teams* will allow you to give agility and speed to your team to achieve the purpose and objectives of the project.

Below are the most relevant steps that will allow you to create a high-performing *Autonomous Team* for your project.

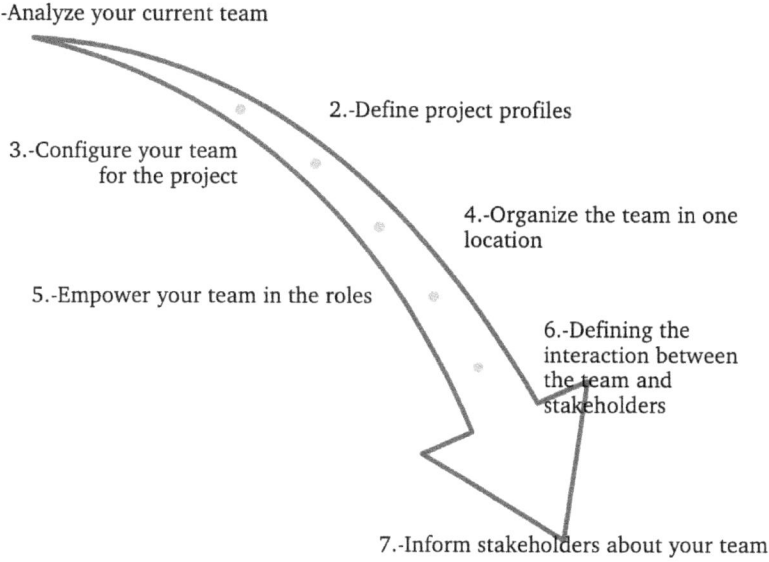

1.-Analyze your current team

2.-Define project profiles

3.-Configure your team
for the project

4.-Organize the team in one
location

5.-Empower your team in the roles

6.-Defining the
interaction between
the team and
stakeholders

7.-Inform stakeholders about your team

Figure 12 Steps to create Autonomous Teams

Step 1: Analyze your current team.

The first step for the creation of your *Autonomous Team* is to analyze the different profiles and maturity of the current team, to assume the new responsibilities and competencies during the development of the project.

If your analysis shows that there is not the necessary maturity in your team, nor can the existing profiles lead this new stage, you will have to have an external leader who can help the transition and management of the team, in addition to hiring new profiles that can count on these skills and abilities to assume the new challenges of being an autonomous team.

Step 2: Define project profiles.

Analyze the new profiles needed to tackle your project, for example: Analysts, Developers, Software Architects, Network Architects, Consultants, etc.

First, search internally in your team, if you have the required profiles to incorporate them into your *Autonomous Team*.

Second, determine whether current profiles can meet challenges such as:

- o Understand the purpose of the project beyond their own scope of work and the impact on the business.
- o Understand why they are performing their tasks and the impact on the development of the project.
- o To be able to detect the value of their deliverables and the quality of these to the business.
- o Generate assertive communication in order to eliminate obstacles and clarify doubts quickly.
- o Be able to help the client prioritize the impact of each deliverable for the project and its value in the business.
- o Proactivity and self-management in advancing in the assigned tasks without the need for direct control to complete them.

Step 3: Set up your project team.

Organize your team into cells of no more than 6 to 8 individuals, each comprising multiple roles such as developers, consultants, analysts, and leaders. This arrangement enables each team member to address a broader range of issues, provides greater **functional coverage**, and allows for the autonomous integration of additional functions within the team.

Each *cell* will be able to develop one or more deliverables from conception to customer approval, streamlining the entire release process for delivery of results.

Step 4: Arrange your team in one location.

Co-locating your team or cells will enhance communication and synergy among members, preventing the formation of *silos and isolated information* within the team.

If the team or cells cannot be located in the same physical space and must operate remotely, it is essential to establish clear agreements for their interaction with each other and their environment:

- The mechanisms to maintain fluid communication between members,
- Definition of communication channels and team meetings
- Dispute resolution procedure to move forward.

Step 5: Empower your team.

Empowering your team implies a great responsibility and trust in the team, in the sense, that it can make decisions autonomously and in a timely manner, without having to climb the hierarchical levels to advance. It is important that, depending on the maturity and evolution of the team, they can make a set of decisions autonomously.

To implement the above, you must define a list of topics in which the team can make decisions autonomously, for example:

- o Data and database design aspects.
- o Technical aspects of integration and exchange of information between products.
- o Improvements in internal processes for the deployment of the product in an accelerated way.
- o Correction of errors immediately, from the moment errors are detected, without waiting for them to be informed by the client.
- o Assignment and prioritization of internal team tasks.
- o Define in which meetings with the client, can make decisions and / or participate autonomously.
- o Other aspects of the project.

Step 6: Define the team-stakeholders interaction.

The important thing in this step is that you clearly define how the relationship of the team, and the stakeholders will be.

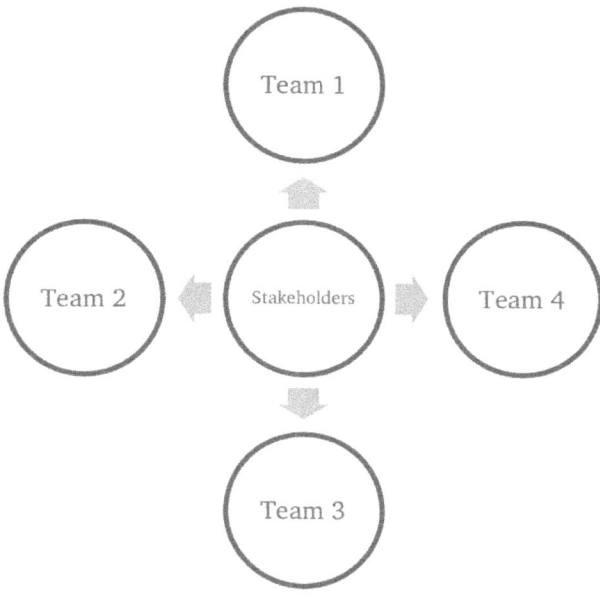

Figure 13 Team-Client Interaction

Step 7: Inform your Stakeholders.

At the commencement of the project or iteration, it is important to inform the client of the responsibilities and authority of the autonomous team to facilitate efficient communication and decision-making.

Delocalized and Remote Autonomous Teams

When your project has delocalized and remote teams, it is important to consider the following alternatives:

Delocalized Cell Team

It organizes remote teams across different locations based on their profiles, ensuring that each remote team functions as an autonomous unit with diverse skill sets. This arrangement allows them to address deliverables independently, minimizing the need for extensive support from other teams, particularly when there is a significant time difference of more than 5 hours.

Teams by Specialty Delocalized

If your remote teams are distributed by specialty in the different locations (*for example: Developers in one site and Analysts in another*) to be able to release deliverables in an agile way, you will have to consider some important actions, which must be transformed into team habits in the short term to be successful:

o   Report the objectives of the week, clearly pointing out the deliverables to be addressed. This activity must be informed through face-to-face meetings or via videoconferences where the whole team participates.

o   Redouble communication in both directions, with daily meetings at the beginning of each day to align the different teams and focus on the specific tasks in which they must work together.

- o Use instant messaging tools for real-time communication between team members.

- o A single email for project topics, in order to keep all teams informed of the latest news of the project.

Interaction of Delocalized and Remote Teams

Below is the interaction of offshored and remote teams working with support cells.

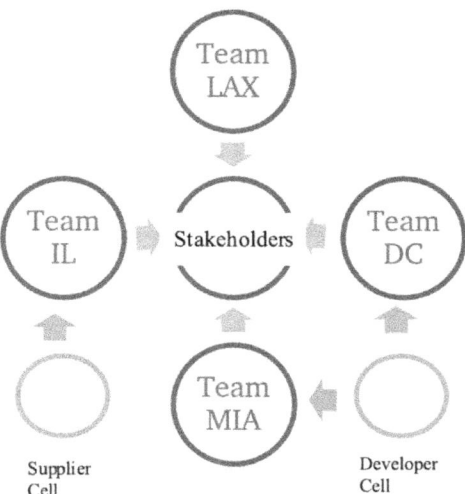

Figure 14 Team interaction with support cells

Autonomous teams with suppliers

In the case of teams with external suppliers, we will analyze two possible scenarios:

Supplier as part of the team

In situations where the supplier's team is integrated as an internal team member and operates from the same location, the same procedures outlined above for creating autonomous teams can be applied, even when suppliers are involved.

Supplier as an external cell to the team

When the supplier works separately from the team, as an *external support cell* to the team, this corresponds to an external service to your organization, so the first thing is to clearly define the functions that the supplier will perform, second the required deliverables, and service levels in variables of response time, quality and values per hour or task of the deliverables, In the same way that the internal activities that the team can develop autonomously were defined.

With this specific separation of duties, you will have a new autonomous work cell as part of your team and that will deliver specific deliverables to meet your objectives.

Interaction of Autonomous Teams with Support Cells

Here's how your team would interact with external support cells.

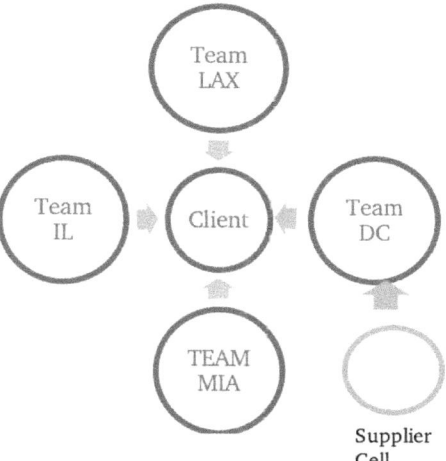

Figure 15 Autonomous Team with Cells

Cooperation in Autonomous Teams

Cooperation in autonomous teams is a key feature for the success of the project and obtaining the results expected by the client.

Some of the benefits of cooperation in *autonomous teams* are as follows:

- o Build relationships within the team.
- o Adds value to the team and client.
- o It generates trust within the team and reinforces its organizational structure.
- o Expand team knowledge to drive results.
- o Generate better results for the client.
- o It generates identity of the team with the value it delivers.
- o Increases the productivity and agility of the entire team.

Case Study

RET INC

Continuing with the *PROSPECT project,* the company *RET INC* has finally determined that the ideal team to start with the project will be formed as follows:

- o Senior project leader (Greck Gast) in Chicago.
- o Senior Team of Analysts (Jean Ro, Robert Hlis) in Chicago.
- o Senior team of developers (Hortar Ju, Hans Ils) in Atlanta.
- o It has been defined that there will be an external provider to address developments from Chicago.

Team structure

The structure of the project team will be defined as follows:

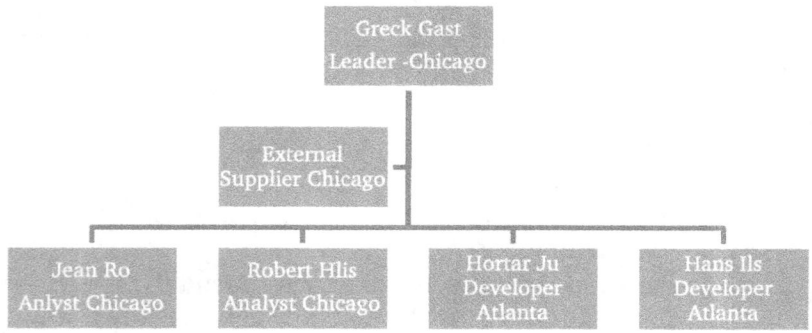

Figure 16 Team structure

Team interaction

The interaction of the team with its supplier and the client will be defined as follows:

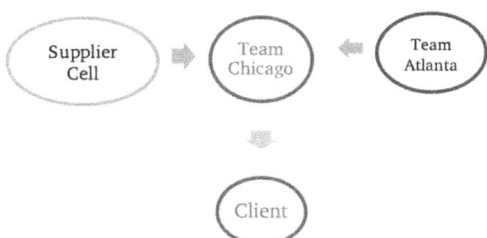

Figure 17 Team with supplier

Note

One characteristic of autonomous teams is that not all cells are required to interact directly with the client. In our specific case, only the Chicago team engages directly with the client.

STRATEGY 3: MANAGE YOUR CLIENT

"Guide your client, as a guide guides the way."

This strategy aims to manage the client proactively, ensuring that value is delivered in every aspect of the management process.

Based on the analysis of the client's scope, management, and communication, you can now define the management style for engaging with the client throughout the project's development.

Management analysis

To initiate the management analysis, we will focus on key topics that directly impact client management.

Generally, individuals and teams are significantly influenced by their culture, personality, and working environment. Therefore, it is crucial to analyze these cultural influences when making decisions, particularly in specific situations that may arise during the project.

Cultural Considerations in Management

One critical aspect to address before undertaking any management activities is the cultural considerations of the clients, as these can directly impact project management.

As a definition, we will point out that *culture* allows people to share common motivations, experiences, and events. Which are transmitted from generation to generation. This marks a way of seeing life, expressing oneself and communicating in all areas of it.

Given the above, in practice, you will not be able to manage a client from *Asia in the same way* as another from *Latin America*, in terms of languages, idioms and ways of approaching management and problems during the project.

For this reason, it is necessary to know some cultural aspects of your client before starting the project.

Let's look at the case of project meetings for two different clients, to see some considerations that are important to analyze.

Asian Client:

- o When introducing yourself, you should prioritize greeting the senior-most and most experienced individual in the organization.

- o The greetings should be made through a bow, bowing your head, and looking into their eyes.

- o When you deliver your business card, do it with both hands, it is a sign of good manners and respect for the recipient.

- o Seating arrangements will be directed by the senior-most client representative, after which the remaining team members may take their seats.

o The meeting will start at the stipulated time and a previously established agenda will be followed.

o No jokes or extensive conversations of topics that are not on the agenda are allowed.

Latin Client:

o The greeting in general is shaking the hand of the client, as a sign of respect.

o When introducing yourself to the client, there may be a more informal conversation.

o Prior to the meeting, attendees could talk about topics such as: weather, sports, family, etc.

o The meeting begins when all the members have arrived in the room.

o There may be consultations outside the agenda, which must be managed appropriately.

Each of the points previously discussed demonstrate the cultural diversity that you can find in your project. So, it is important to analyze these cultural differences, otherwise, you will not achieve an effective *management of your project*.

Here are some recommendations for managing meetings with clients from diverse cultures in your project.

Recommendations for managing with diverse cultures.

Some recommendations for dealing with different cultures are the following:

- o Introduce yourself to clients and stakeholders in a formal way, validate the type of greeting, which is culturally correct for your client.

- o Arrive at the meeting location at least 15 to 20 minutes early, to prepare all the necessary Team to start.

- o If the meeting is remote, use a camera to present yourself visually, this generates a closer connection with your client.

- o Use clear, formal language to explain each of the topics. Do not use local idioms, slang or dialects.

- o Watch your nonverbal language and mannerisms during meetings, to avoid misinterpretations.

- o Be clear and concise in explaining each of the topics being addressed.

- o Refrain from making jokes during the meeting, as this may distract attendees from the topics to be discussed.

- o Maintain emotional self-control to manage clients from different cultures, since this type of management is associated with a high level of stress.

- o Prepare each client management activity in advance.

o You should be prepared to consider that some management may take more time and effort, due to the cultural difference and organizational structure of your client.

o Previously analyze their interests and needs before proceeding with any management with your client.

o Visualize your client as a partner to achieve project success.

o It is important that during the negotiations you achieve credibility, empathy and detect the real needs and pains of your client.

o Use excellent communication during meetings and negotiations with your client.

Negotiations with your client

Conducting a correct *negotiation with your client* is a competence that you must master for success in the management of your project.

The relevant steps that you should consider when carrying out a negotiation with your client are:

o Define the objectives you are looking to achieve and a margin in which you would be satisfied to close the agreement (maximum and minimum).

o Look for common interests that allow you to generate bridges of approach to reach an agreement with your client.

o Problems are resolved through open dialogue and willingness on both sides. Be empathetic to your client.

o Put yourself for a minute in the position of your client and think about what he seeks to achieve with the agreement, what benefits he expects from it and what would be his acceptable margins to close the agreement.

o Analyze how your client approaches negotiations, this involves listening and understanding the way and the way they communicate, in addition to their non-verbal language.

o Determine what your client's interests, options, and alternatives are to move forward, rather than their current position in negotiations.

o It aims to broaden the range of solution options by presenting multiple alternatives to the problem, incorporating new precedents or benefits to facilitate a revised negotiation scenario.

o Strive to achieve a win-win agreement where both parties perceive that they have received adequate benefits from the arrangement.

o Self-control in negotiation and management are key to reaching a good agreement.

o When you get the verbal agreement, make a document to record the specific terms of it and sign it with your client.

List of procedures

Below, we provide a detailed list of procedures that you must undertake with your client:

Contract negotiation

Regarding the *management of the contract* with your client, it is important that you consider the following points:

o   Analyze which are the aspects that you are interested in leaving in the contract and are important for your project.

o   You detail the key clauses that you are interested in including:

- Obligations for the client.
- Obligations for the supplier.
- Type of contract modality:
  - Fixed price.
  - Time and material. (*a price per hour and the material used is fixed*)
  - Subscription model - SaaS
  - Other.
- Contractual and employment relationship between the parties.
- Confidentiality clauses.
- Data security and ownership
- Service level agreement SLA
- Metrics and/or environmental conditions of the product.
- Project deadlines and milestones.
- Liability and Warranties
- Expected deliverables. (*if necessary, attach technical specifications*)
- Payment method and associated charges.
- Arbitrations and mediation in case of disputes.
- Exit clauses and early termination of the contract.
- Other.

o Conducts contract review meetings, depending on the evolution of contract versions. At least once a week to review progress.

o It is important that after each contract review session, you leave in writing and formalized in the same contract document, the adjustments and terms agreed or added.

Note

Never start your project, if there is no agreement about the clauses of the contract and it is not signed, since it increases the risk of project management.

Kick-off meeting

In this initial meeting, you must be clear about the approach you will use, based on this approach you will define the essential points you want to present for the start of the project.

It is important in this initial meeting; you take a leadership role for the control of the agenda and times stipulated in each of the topics that you will present to the client and the stakeholders of the project.

Here are two approaches you can use for the initial customer meeting:

o Focus 1: Executive Vision (20-30 min)

- Purpose of the project, as an overview of the product.
- Presentation of the Client and Team scopes.
- Presentation of the field of Communication and Management.
- Words of the sponsor of the project.
- Next actions.
- Doubts about the project.

- Closure of the meeting.
- Sending minutes.

o Focus 2: Detailed Overview (45 min - 1 hour)

- Purpose of the project, as a vision of the expected product.
- Presentation of the Client and Team scopes.
- Presentation of the field of Communication and Management.
- Words of the sponsor of the project.
- Vision of the plan and important dates of the deliverables.
- Initial risks.
- Next actions.
- Doubts about the project.
- Closure of the meeting.
- Sending minutes.

Note

Previously, the Project Leader and the Sponsor have agreed during the Communications and Management scope, the date and place of the meeting.

It is normal that, at the beginning of an iteration or start of the project, the client does not have clarity regarding what he needs in detail, so your role will be fundamental to guide the client in achieving what he really needs to define to solve his problem.

Definition of Requirements and Changes

For meetings and sessions to take requirements and changes, you must previously define:

- o The type of requirement or change you want to define.
- o The profile of the customer to be involved.
- o The team members you want to participate.
- o Define the sections of the requirements or change document.

After finishing the meeting to take requirements or changes, send the client the document with the described detail to approve the requirement.

After all meetings of requirements or changes, ask the client to sign the final document to manage the development of the deliverables.

Note

For more information on Requirements, review the chapter on the Effective Requirements Strategy.

Control and monitoring of deliverables

It is important that you plan the management of the topics to be discussed in this type of meetings with the client, which should include:

- o Current status of deliverables under development.
- o Closing dates for validation.
- o Problems and risks detected with the client.
- o Clarification of doubts to advance in the deliverables.
- o Prioritization of deliverables for the next meeting.

Additionally, and as part of the client's management, it is important that in these meetings you manage the client's expectations about the deliverables and thus avoid last-minute surprises, which may cause unnecessary conflicts.

Review and acceptance of deliverables

For this process, it is important to previously manage:

- o   The test plan of the deliverable.
- o   Prepare the environment where the review will take place.
- o   Have at least one team and customer representative.
- o   Prepare a document of Acceptance of the Deliverable.
- o   Review the deliverable acceptance document for the customer to accept the deliverable for signature.

Note

For more information on Deliverables review the chapter on the Strategy "Search for Results".

Customer Training

For this management it is necessary:

- o   Agree with the client the dates and time allocated for the training.

- o   Define with the client, if there will be a leading user of the system, to differentiate the type of training that will be given to this user during the process.

- o   Define a capitation plan for the deliverables, according to the dates and time allocated for the process.

o   Prepare manuals for the use and operation of the product.

o   Manage with the client, before giving the training, the approval of the training plan.

o   Manage with the client, the signature of the realization of the training, once completed.

Project Closure and Contract

For this management it is necessary:

o   Prepare a project closing meeting with all project stakeholders.

o   Prepare a document of acceptance and closing of all the clauses of the contract:

- Where accepted deliverables are mentioned.
- The documentation delivered.
- In addition to mentioning, that everything stipulated in the contract has been complied with.

o   In the scenario, that the client accepts *the project closing document*, proceed to sign the document in the same meeting as a sign of acceptance and closing of the contract.

Offshored Customers

In projects where your client is delocalized and the interaction is only remote, the complexity of these projects is quite high. All procedures, such as: resolving pending issues, communications and conflict resolution are much slower and more complex. All these factors increase the complexity of your project as we have previously mentioned.

Our recommendation for this type of project is to apply some of the following actions that help you reduce the complexity of the project:

o Prior to the meeting, send the agenda of the topics to be discussed.

o Use video conferencing software during the meeting.

o Share the screen with the topics you are reviewing, so that both parties see the same thing.

o Send a minute after the meeting.

o Increase the periodicity of meetings and sessions to clarify doubts with the client.

Note

It is highly recommended to conduct the requirements definition process in person. This approach fosters trust, enhances scope clarity, and reduces the need for subsequent rework in the project.

Case Study

RET INC

Continuing with the *PROSPECT* project, the following events have occurred:

o The contract has been signed.
o Customer and internal teams have been defined.
o Project dates and deadlines have been agreed.

The next step is to prepare the initial project meeting.

Efforts

The steps to consider for the initial meeting are as follows:

- o Manage with the client.

  - The agreed day for the meeting will be Monday, February 23.
  - The level of depth of the meeting will be detailed.

- o Manage with the team.

  - The role of each team member.

  - The autonomy functions of the team during the project.

  - Prior review of the general scope of the project and the initial risks that are visualized.

- o Preparation of the meeting

  - Sending of appointment of the meeting with the topics to be discussed.

  - Validate the communication system used for the meeting with the relocated members.

- o Post-meeting

    - Sending minutes with the topics discussed and agreements.

    - Next actions of the team and client.

STRATEGY 4: COMMUNICATE 360

"If you don't communicate it, it doesn't exist in your project."

During and after the pandemic, the strategy of communicating properly during the development of the project became a key and fundamental strategy when managing and controlling your project. Communication is the foundation of all successful project management.

Considerations for effective communication

There are some essential considerations for effective communication, among which we can mention the following:

o   What do you want to communicate?

   ▪   A concept, desire, order, a story, etc.

o   Who do you want to communicate to?

   ▪   Team, customer, supplier, etc.

o   How?

   ▪   You define how you are going to communicate the message.

   ▪   Depending on what you want to communicate and to whom you should send the message, you define the means by which you want to transmit it, for example written, oral, remote, face-to-face, other communication.

o   For what?

- You must define the objective sought with the message, but you have clarity for you to communicate, you will not know if the process was successful.

Effectiveness in forms of communication

Communication can be more effective depending on the way you communicate, for example: an in-person meeting is more effective than an email, since in the first you can communicate verbally and non-verbally.

Below, we show the effectiveness depending on the medium you use to communicate:

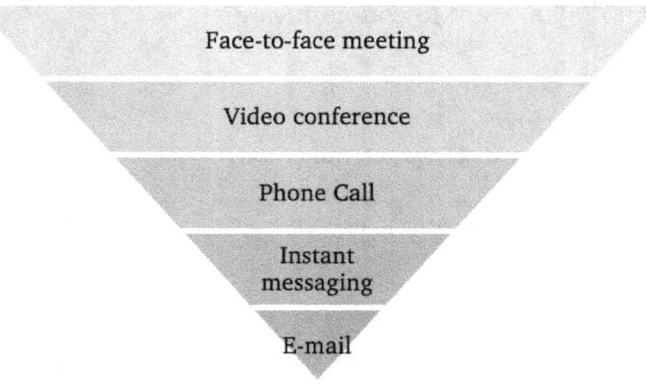

Figure 18 Effectiveness depending on the means of communication.

Communication problems

The lack of communication in your project can generate the following problems:

- Errors in the interpretation of customer requirements.

- Lack of understanding of the deliverable to be released within the team.

- The result of the deliverable is not in accordance with what the customer expected.

- Team members in other locations are not clear about the tasks they need to tackle.

- The team is not clear about the priority of its tasks.

- The customer does not know when they will receive the first deliverable.

- The requirement has not been clearly defined for development and should be reviewed again with the client.

Each of these problems can generate a significant impact on the project, in terms of time, quality and additional expenses for rework and others.

Communicate based on the profile.

In your project you will have different types of communication depending on the profile of your client or team, for example:

o The Millennial Generation (1981-1993) communicates natively via instant messaging and social networks.

o Generation X (1969-1980) is generally comfortable communicating via email or a phone call.

o Baby Boomers (1949-1968) generally prefer face-to-face, telephone and paper communication.

Note

These statements are generalist, so, depending on your project and the profiles of your client and team, you must validate why you will communicate more effectively with each of them.

Recommendations for communicating in different situations.

Here are some recommendations to improve the way of communicating during the development of the project:

o To communicate, create a climate of collaboration with the team and the client.

o Communicate good news and problems at the right time, do not postpone communication.

o Be honest in communicating the news.

o Get ahead of situations and communicate in a timely manner, as long as you communicate stay in flames.

o Maintain the same formality to communicate the various situations of the project.

o Standardize the way you will communicate during the development of the project.

Communication during Meetings

For meetings it is important to:

- Determine the ideal participants for the meeting.

- Send the agenda of the topics to be discussed.

- Sends the appointment in the recipient's time zone.

- Introduce yourself to the client in a formal way and then introduce your team if it's the first meeting.

- If the meeting is remote with clients or offshored teams, prepare messaging communication systems and software that allow you to share the screen to review the topics.

- Start the meeting by reviewing the topics to be discussed.

- Lead the control of the agenda, starting with the first item on the agenda.

- If any topic extends beyond what is stipulated, inform that it will be discussed in a separate working meeting, to advance to the next point.

- Do not use idioms, only formal and precise language in explaining each of the topics that are being addressed.

- Send the minutes with the topics discussed and the agreements reviewed.

Key actions to streamline the communication of remote and offshored teams.

o   Communication interactively through chat, instant messaging, and video calls.

o   Create email groups for the team and the client, where project information can be shared quickly and clearly.

o   Daily work meetings to validate pending and doubts of each member of the team.

o   Communication must be clear, fluid, and fast to avoid misinterpretations by the recipients of the message.

Note

If it is necessary to repeat or clarify the message more than once through different communication channels, it is always appropriate to do so.

The messages sent can generate erroneous interpretations of the information depending on the channel through which they are transmitted.

If you maintain a high level of communication in remote teams, great results can be achieved.

Communication and Leadership

There are several types of leadership that can help improve communication, among which stand out the leaderships that actively participate with the team and give them the necessary prominence to develop.

Below, we mention some leaderships that can help the way of communicating during the project.

- Participatory Leadership

    - They integrate the team in decision making, this is important to increase the communication and experience of the team in decision making.

- People-Oriented Leadership

    - Empower your team and their collaboration, increasing communication with the team.

- Laissez-faire leadership

    - They give autonomy to the team in making their decisions, but control and correct based on their results, communicating to the team actively.

    - This is relevant for autonomous teams.

Communicate 360

As previously, we defined the scope of *Management and Communication (Strategy 1)*, now is the time to define the Strategy of communicating 360.

This strategy consists of: *"Communicating effectively and concurrently to the Client and Team areas, maintaining an open channel of communication throughout the development of the project"*.

Communicating to these two areas covers all aspects of project communication. If any of these areas does not have adequate and timely information, you could enter a communication crisis, such as those discussed above.

Never leave information gaps in any of the areas of the project since this asymmetry of information can generate delays and problems.

When you communicate 360, the important thing is that you consider the following awards:

o   The communicated message reaches its receivers regardless of the way you have chosen to issue it.

o   The message issued is clear.

o   The receiver has understood the message correctly.

Meetings for 360 communication

To reinforce this strategy, it is important to define at the beginning of your project a set of meetings that allow you to support 360 communication. This set will give you a communication framework and maintain an *open channel of communication* for all areas of the project.

Next, we point out the set of meetings that can help you maintain an open communication channel for the different areas.

Client Scope

o   Requirements and Changes sessions.
o   Control and Progress Meetings.
o   Validation meetings of deliverables and functionalities.
o   Iterations and project closing meetings.

Team Scope

o Internal and client work meetings.
o Internal progress meetings.
o Risk meetings.

Product Scope

o Meetings to define Deliverables and tasks.

Communicate 360 with remote teams and clients.

Effective communication in offshored teams and customers is key to the success of your project.

Advantages

One of the advantages of remote teams with different time zones is the ability to use these jet lags to complement jobs across each of the computers. For example: team *A* of developers in the first locality sends a version of a deliverable, so that *team B* in the other locality can test on their schedule. Once the tests are finished, *team B* sends the observations and pending tasks to *team A* for the next day.

Disadvantages

One of the drawbacks of very pronounced jet lags is the coordination to hold meetings with all teams. The important thing at this point is to be flexible and define at the beginning of the project the days and times of the coordination meetings of the teams, determining a schedule where everyone can participate with a little flexibility, either early, at noon or very late.

STRATEGY 5: *EFFECTIVE REQUIREMENTS*

"If you can't define it, you can't create it."

One of the key strategies to correctly define the scope of the project or an iteration of iteration is to define requirements clearly and effectively at the right time.

This strategy proposes a series of steps and recommendations to effectively define your requirements.

Let's start by defining what a requirement is, and then review the strategies to address them.

What is a requirement?

In general terms, a requirement is a specification of the product to be developed, provides a solution to a business need. Each of the requirements defined during the project are consolidated in the *Requirements Specification Document*.

In the following sections, we will review in detail, the content of a requirement and how they should be defined for proper use in later stages of the project.

Cycle for requirements management

The cycle to manage requirements effectively has the following steps:

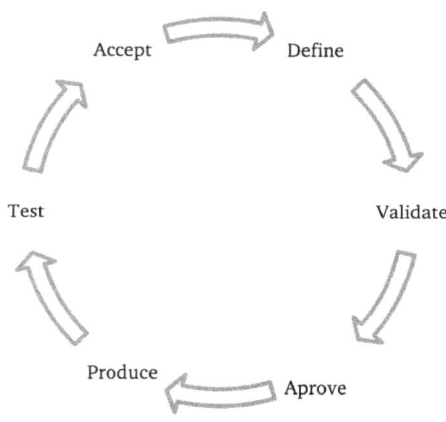

Figure 19 Requirements Management Cycle

Define

At this stage, the problem to be addressed is defined. To facilitate this process, requirement sessions are scheduled, which will be detailed subsequently.

Validate

At this point, the *requirements specification document* is validated and adjusted to correctly define all cases and aspects of the problem to be solved.

Approve

At this stage the customer must accept the *requirements specification document*.

Produce

After all the previous steps, we proceed to develop or produce the deliverable that will respond to the requirement.

Test

First the team and then the customer perform the functional tests of the deliverable, verifying that it complies with everything detailed in the *approved requirements specification document*.

Accept

The client, after testing and validating the deliverable, formalizes the approval of the deliverable, signing the acceptance document.

Management cycle for different types of projects

For both predictive and adaptive projects, it is important to define and detail the requirements at the beginning of the project or each iteration, in order to maintain the consistency of the deliverables and prevent the client from changing the requirement during its development.

If later, after the development is finished, the client needs to adjust or change the requirement, a change is defined, which must enter the normal cycle of *requirements management* to be produced or developed.

If you do not have an order at this point, you could end, in an extreme scenario, changing the requirement as many times as the client wants, without ever finalizing its delivery and subsequent approval.

Steps for requirements management

Here are the steps to prepare for the *requirements* sessions:

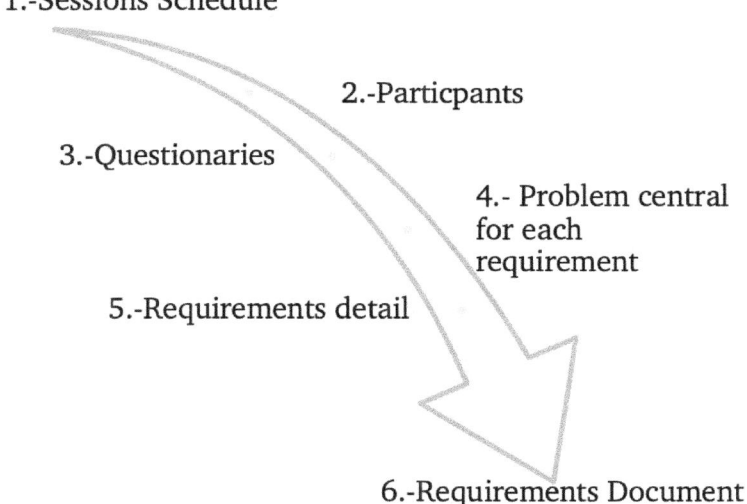

Figure 20 Steps of Requirements

Step 1: Session Schedule

Defines a schedule of sessions, to perform the analysis of needs, both functional and non-functional with the various clients of the project.

Step 2: Session Participants

Determine the participants in each of the requirements definition sessions, based on the information previously defined in your customer scope.

- o Identifies the source users of the requirements.
- o Identifies the different classes of users for the product and/or service.
- o Identify the representative user for each requirement.

Step 3: Define Questionnaires for requirements.

For each requirements session, you will need to prepare a questionnaire with the questions you want to answer during the session, for example:

- o What do you want to solve?
- o What does it need to be solved for?
- o Where does the problem occur? In what context?
- o How many actors or parties are involved?
- o When does it happen?

○ When do I need to solve it?
○ What information is needed as input, during troubleshooting and output?

Note

Upon completing the questionnaire, you will gain insight into the significance and priority of the defined requirements.

Step 3: Define the Core Problem

In the first analysis session, your objective will be to define and understand the central problem you are looking to solve.

This session will give you an overview of the problem to be solved.
You should understand the process and context in which the problem was detected. Next, understand where the problem has been detected, and you must also capture the activities and who performs them today.

Step 5: Requirements Detail

In the following sessions, each of the requirements will be defined in detail, based on the defined questionnaires, and using the tools mentioned below:

Each session should address specific topics, in time intervals of 1 hour, a break time of 15 min and then, if necessary, continue a session of 1 hour.

Step 6: Requirements Document

Document each requirement, and after each session, send the collected document to the client for review and approval, indicating the date of creation, date of approval, client that has approved, priority and area of interest to which it belongs, among others, as shown in the following tab of the requirement:

Number : *1*
Requirement Name : *Customer entry screens*
Area of interest : *Business*
Priority : *Stocking*
Risk detected : *Multiple clients with different.*
screen layouts, can generate.
Approval delay

Creation date: *:17 /03/2019*
Date Modified *:18/03/2019.*
Approval Date *:19/03/2019*
Author : *Roger Thol*
Approved by : Alan HUl
Approving firm : Alan HUl
Requirement Detail :

Note

Do not move forward with the following requirement, if the definition is not clear, doubts remain, or the document is not approved in its entirety.

If necessary, define a special session for adjustments, revision, and clarification of the document.

Tools for defining a requirement.

The tools you can use in each of requirement's sessions are varied and will depend on how effective they can be for you, below, we detail each of these:

Questionnaires and surveys.

- o The questionnaires are key to the definition of the requirement, since they allow you to understand the problem to be solved, through the answers obtained.

- o Surveys can also be useful when you need answers from a customer who is not physically in the same location or validate a group of users' perception of a particular topic.

Interviews.

- o This tool seeks that during a session of taking requirements you can conduct an interview in person with your client, applying a questionnaire, which allows you to know first-hand the answers to document the requirement.

Focus Group

- o In the context of the project, it consists of bringing together a group of clients, to deal with one or more requirements so that you can collect different opinions on the same topic.

Brainstorming and mind maps

- o It consists of arranging a set of ideas on a specific topic and analyzing their relationships.

- o This tool can be used mainly with the project team to find alternatives on how to develop or solve a project requirement.

Analysis of documentation and business processes.

- o This tool allows you to review previous documentation on existing processes and systems, helping you understand the context and define the requirement.

Reverse engineering of existing products.

- o This consists of taking the existing product and documenting each of its parts and functionalities, to be improved or replaced by another.

- o Reverse engineering is useful when you have to replace a system that is obsolete, and you need to analyze each of the parts or functionalities that the current system performs.

Prepare Prototypes

- o This tool consists of visually showing through a mockup the requirement or part of it.

- o This tool is useful for requirements with many visual functionalities since you can then attach the selected prototype to the requirement document as part of the definition.

Use Case Diagrams

o   The use case diagram allows you to give an overview of the activities that users perform with the system for a certain functionality or requirement.

Defining a Requirement

During the requirements sessions, it is important to have reviewed each of the questions indicated below, so that the requirement has been clearly defined:

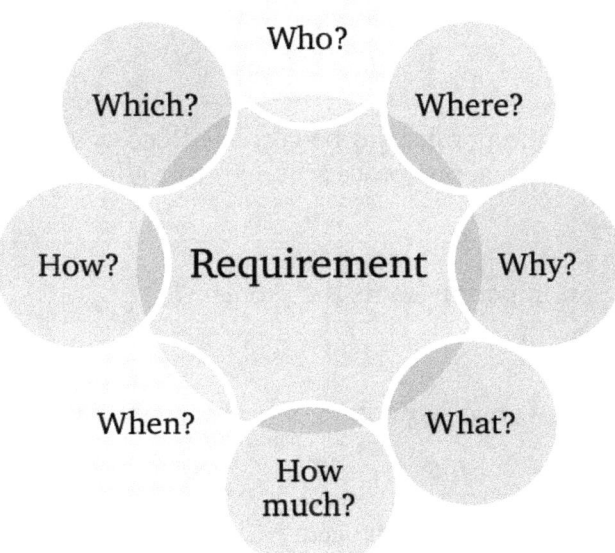

Figure 21 Quality of Requirements

What?

- o   What problem needs to be solved?
- o   What impact does this problem have on other areas, products, or the organization?
- o   What will happen when it is resolved?
- o   What could happen if it is not resolved? –Aftermath-
- o   What impact does it have on stakeholders?
  - •   For example: employees, suppliers, customers, users, the business, etc.
- o   What limits does the problem have in terms of organization, workflow, geography, and segments?

Who?

- o   Who is affected by this problem? Groups and stakeholders, consumers, or organizations.

Where?

- o   Where is the problem to be corrected occurring? only in one place, one area, one process, in many products, etc.

Why?

- o   Why is it important to fix the problem?

How much?

- o   How many areas or departments are involved?

When?

- o   When does the problem occur?
- o   When does it need to be resolved?

How?

- o How are activities performed today to support this problem?
- o How do you expect to use the results when the problem is solved?
- o How should the result of the Problem be displayed?

Which?

- o What are the Conditions of satisfaction? to validate the solution meet the requirement.

Note

Conditions of Satisfaction, also known as acceptance criteria, are crucial to document as they define what must be delivered and specify the conditions under which the requirements will be deemed acceptable.

Skills to define a requirement.

The team member assigned to define the requirement must possess the requisite skills to execute the process effectively:

Figure 22 Skills to define a Requirement.

1. Active listening (*understanding*).
2. Facilitator for the client.
3. Analytical to understand the contribution of the requirement to the business.
4. Write the requirement correctly (*see section on quality requirements*).
5. Observer.
6. Organized to achieve focus and deploy the ideas of the requirement.
7. Creative to propose new alternatives to the requirement.

Quality of Requirements

When you have defined a requirement, it is important that you know if it will have the expected quality for its development and subsequent validation by the client, for which it is important that you validate the following points:

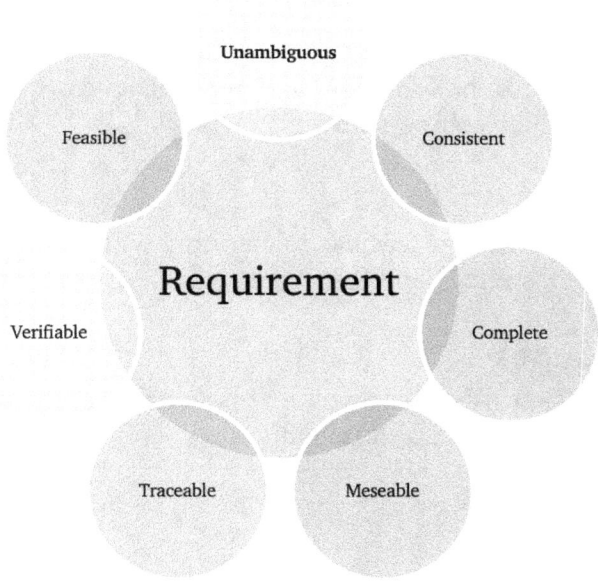

Figure 23 Quality of Requirements

Verifiable
   o   It must allow you to test and verify your content.

Traceable

o   It must present a single requirement and not be mixed with another.

Measurable

o   It must allow you to validate your output against the defined acceptance criteria.

Complete

o   It must have all the necessary detail for its design and development.

Consistent

o   It must be consistent and not contradictory with respect to other requirements.

Feasible

o   It must be feasible to perform with respect to existing technology and resources.

Unambiguous

o   It must be clear and not allow misinterpretations of both the team and the client.

Requirements Prioritization

A quick way to prioritize *requirements* is the scale of importance and value it has to the *customer's* business.

Requirement	Dimension / Business
Requirement 1	High
Requirement 2	Casualty

It is also possible to do so, depending on the area of interest in which the *requirement is located* and prioritize it from the greatest to the least impact in each of its areas of interest. As the following table shows:

Requirement	Dimension			
	Business	Legal & Environmental	Critical Process	Technological
Requirement 1	High			High
Requirement 2		Medium		

Case Study

RET INC

The company *RET INC* has determined to start the sessions of taking requirements of the *PROSPECT product*, for which it has defined:

Schedule

Initial session	: 10 March	10-12 AM Room 1
Session 1	: 12 March	10-12 AM Room 2
Session 2	: 17 March	10-12 AM Room 3
Session 3	: 18 March	10-12 AM Room 1
Session 4	: 19 March	10-12 AM Room 2
Session 5	: 20 March	10-12 AM Room 4
Session 6	: 21 March	10-12 AM Room 1

Participants

Initial session	All the team Clients, including the sponsor.
Session 1	Greck Gast Leader -Chicago Jean Ro Analyst- Chicago Robert Hlis Analyst- Chicago Alexander Fle - Customer Leader Artur Klo - User Bennet Hiu – User
Session 2	Jean Ro Analyst- Chicago Artur Klo - User
Session 3	Bennet Hiu – User Robert Hlis Analyst- Chicago

Session 4	Jean Ro Analyst- Chicago
	Artur Klo - User
Session 5	Bennet Hiu – User
	Robert Hlis Analyst- Chicago
Session 6	Greck Gast Leader -Chicago
	Jean Ro Analyst- Chicago
	Robert Hlis Analyst- Chicago
	Alexander Fle - Customer Leader
	Artur Klo - User
	Bennet Hiu – User

Questionnaires

We will prepare a questionnaire to determine the requirement called: "Conversion Effectiveness Calculation."

What?
- What do you want to calculate in conversion effectiveness?
- To what extent should the result of the conversion effectiveness calculation be presented?
- What input parameters should be used for the calculation?
- Can there be negative parameters and results?
- What impact does this calculation have for the commercial department and the other areas of the company?
- What time limit is there for the system to calculate conversion effectiveness?
- What values (maximum and minimum) should be obtained from the calculation?

Who?
- Who is the user in charge of entering and validating the input parameters?

o Who is the user in charge of calculating the effectiveness value today?
o Who are the users who will be able to view and use the calculation result?

How?

o How are activities to calculate the effectiveness value done manually today?
o How should the result of the Calculation be displayed?

Where?

o Where are these parameters arranged to be entered into the new product?
o Where should the input data be extracted from?
o Where should the calculation results be displayed on a website, on a phone, etc.?

Why?

o Why is it important to have the calculation in the expected time?

How much?

o How many areas or departments will use the calculation?
o How many external systems will it be necessary to integrate with to deploy the calculation?
o How often should the input parameters and result be updated?
o How long at most, should an effectiveness estimate be obtained?

When?

o When is it necessary to calculate the conversion effectiveness?

Which?
- o What are the steps that are performed today to perform the effectiveness calculation?
- o What are the satisfaction conditions for the calculation to be accepted?
- o What are the parameters needed for the effectiveness calculation?

From the prepared questionnaire, there may be questions that are repeated, but posed differently, this will allow you to reiterate them in case your client has not understood the first way to pose them.

The questionnaires allow you to answer, if your requirement meets the required quality according to the guidelines, we have previously reviewed.

Central Problem

Based on the initial meeting and subsequent sessions, the primary problem to be addressed by the project has been identified as follows:

The problem arises in the Commercial department every three months, since, in its database, there is a list of prospects that could become future customers and have no way of knowing their conversion level or which of them could become a customer accurately.

This implies that overestimates could be generated in their projected revenues due to the lack of customers, generating significant financial problems during the rest of the year.

The estimated customer conversion cycle is three months, from when a prospect is entered into the database until it becomes a customer.

The Commercial area seeks to have a product for use on the web and mobile phones, which allows you to estimate with 80% to 90% effectiveness, the probability of conversion of prospects into customers, for each quarter of the year.

Requirement Detail

Upon completion of the sessions and questionnaires, the defined requirement is detailed as follows:

1. Conversion Effectiveness Calculation

    1.1. The calculation of effectiveness seeks to solve the problem, of whether a prospect, after three months will become a client of the company. This currently leads to errors in revenue estimates for each quarter of the company.

    1.2. The entry of the information of the prospects is carried out by the four managers of client accounts, during the three months prior to the closure of the system for its calculation.

    1.3. The calculation of effectiveness is carried out by the head of the commercial area, taking the list of prospects entered in the GFT relational database of the commercial system, composed of:
    - The name of the package leaflet (N)
    - Time on the list since entry (T in months)
    - Level of interest (scale of 1 to 3, with 3 being the most interesting) (I)
    - Decision-making power (scale of 1 to 3, with 3 being the most powerful) (P)
    - Budget (B) (scale of 1 to 2, with 2 budget approved and 1 not yet approved) (P).

- o Importance of the product to the company (scale from 1 to 3, with 3 being the most important) (M).
- o Financial situation of the prospectus company (value to calculate Income/Debt) (D).

1.4. The head of the commercial area applies the following formula to calculate the conversion effectiveness value (VEF).

For a prospect N its value would be:

$$FEV = 0.5*(P* M* 0.5 I* D* (3/T)) +0.5* B$$

Calculation example:

$$FEV = 0.5*(P* M* 0.5 I* D* (3/T)) +0.5* B$$

N	T (months)	I	P	B	M	D	FEV	% FEV	Converted Client
Pros1	4	3	2	2	3	1	4,375	2,1 %	No
Pros2	2		3	3	1	1	4	14	6,9%	No

The highest expected value of the calculation will be:

N	T (months)	I	P	B	M	D	FEV	% FEV	Converted Client
Pros1	1		3	3	3	3	10	205	100%	Yes

And the lowest expected value will be:

N	T (months)	I	P	B	M	D	FEV	% FEV	Converted Client
Pros3	12	1	1	1	1	0	0,5	0%	No

1.5. The VEF value is displayed every quarter, in a list that is distributed by mail to the Finance and Commercial department. The mail is issued by the head of the Commercial area.

1.6. It is expected that the system will take less than 5 minutes to calculate a total of 100 prospects, after the calculation is completed send an email with the calculated results of FEV to the Commercial and Financial departments. Additionally, the result must be loaded into the GFT relational database of the Commercial system. And finally, the calculated value must be available on the website designed for this new system and its mobile application.

1.7. The time required for the project from its inception to production will be 3 months.

1.8. The current FEV formula may be modified in order to achieve better adherence to the actual conversion data.

1.9. Historical data from 2013 to date FEV, both FEV estimates and actual conversions, will be available for unit and preliminary testing.

1.10.   To accept the deliverable, the value obtained from the VEF calculation will be compared at the end of the third month with the real customers who have converted. For this analysis, a percentage calculated on 85% of VEF as converted customer will be considered.

Requirements Document

To formalize the requirement previously detailed in the Requirements Specification Document, the following data sheet will be incorporated for its correct registration:

Number	*1*
Required Name:	*Conversion Effectiveness Calculation*
Area of interest:	*Business*
Priority:	*Loud*
Risk detected	*Achieve effectiveness of the calculation.*
Creation date	*17/03/2019*
Date Modified	*18/03/2019.*
Approval Date	*19/03/2019*
Author	Jean Ro Analyst- Chicago
Approved by:	Artur Klo User
Approver Firm	Artur Klo
Requirement Detail	

STRATEGY 6: *DELIVER VALUE*

"Define your goal, then find your way."

To effectively adopt this new strategy, it is essential to transition from simply completing project stages to concentrating on delivering outcomes that provide the maximum value to both the client and the business.

History is replete with projects that have successfully completed their stages yet failed to deliver the value to the business for which they were originally conceived.

This strategy aims to reduce project complexity by breaking down requirements into deliverables, with the goal of maximizing customer value and minimizing expenses related to unnecessary developments.

This strategy introduces the concept of the **Deliverables Tree**, which consists of a structured and prioritized set of deliverables required to develop a product or its components, thereby generating enhanced value for the customer.

What is a deliverable?

A *deliverable* is a functional portion of the product, which corresponds to one or more business requirements.

As characteristics of a *deliverable,* we can point out:

- A *deliverable* must provide specific, measurable value to the customer.

- A *deliverable* is composed of a set of tasks to be developed to complete its development.

- Each *deliverable* has an acceptance criterion defined by the customer and validated by the team.

- Each *deliverable* must contemplate the quality criteria of the requirements it contains.

Create and manage your deliverables tree.

Below, we list the steps you'll need to take to *create and manage your* product-centric deliverable tree.

Figure 24 Deliverables Tree

Step 1: Define the Iteration Cycle and Tree Validations

Define with your client the following time windows:

- The time to release the deliverables tree (*4 to 6 weeks*).

- The period of time to carry out the deliverable control meetings with the client (*Previously defined in the scope of Management and Communications, for example, biweekly*).

- Definition of validation time and acceptance of deliverables with the client (*3 days maximum to validate and give feedback on the deliverable*).

Step 2: Define Deliverables (Results)

- Meet with your client and review the *Approved Requirements Document*. From the review, a list of requirements should come out in descending order, from the most important to the least. Depending on the degree of value it brings to the customer.

- Taking the list of prioritized requirements, each of the deliverables of the product or service will be created.

- Deliverables are defined by grouping one or more requirements that allow defining functionalities that deliver value to the customer.

- Each deliverable must have an acceptance criterion agreed with the client for subsequent review and acceptance. This criterion is obtained by grouping the acceptance criteria for each requirement.

o   Risks that may arise during this process should be documented.

Step 3: Create a Deliverables Tree

o   After defining each of the deliverables, it will be necessary to order and review them together with the client, depending on the greater value they contribute, grouping them in a deliverables *tree*.

o   The creation and validation of each tree deliverable shall not last beyond the total duration of the deliverable tree, (*4 to 6 weeks*) previously defined in Step 1.

o   If the number of deliverables exceeds the agreed time for a tree, these deliverables should be included in the next *deliverable tree*, to be addressed in the next iteration.

o   This grouping of deliverables will provide the client with immediate value in the early stages of the project.

o   It is always important to start with the deliverables that add the most value to the client's business, and then continue with the following deliverables.

Step 4: Define the scope of the Deliverables Tree

o   Once all the previous steps have been completed, the client's acceptance must be requested, for the scope defined in the deliverables tree.

o   Only with this formal acceptance does the development of the tree begin.

Step 5: Create tasks for Deliverables.

- o Start the review with the highest value deliverable.

- o Next, the project team must add the respective tasks, with their priority and initial estimate of effort in hours.

- o Then, to calculate the hour estimate of each deliverable, the estimates of their tasks must be added.

- o The next step is to determine the total time of the *deliverables* tree, through the sum of hours of each of the deliverables and determine if they fit the time window to release the tree (*defined in Step 1*). If it is not possible to include all the deliverables in this iteration, the leftovers will be included in another tree to develop them in the next iteration.

Step 6: Assign task assignees.

- o For this process, each of the tasks must be assigned a person in charge of the project team, indicating duration in hours, a start and end date of this.

Step 7: Run and monitor the Deliverables Tree

- o Biweekly, a control meeting will be held with the client, to review the status of each of the deliverables of the tree, depending on their progress, time and hours invested.

- o The team will be able to meet daily to review the status of each of the tasks in the tree and analyze potential risks.

Step 8: Testing and Correcting the Deliverables Tree

o   During this process, the team will perform a set of tests to the development deliverables, in order to make the necessary adjustments and corrections for their release to the client.

Step 9: Validation and Closure of the Deliverables Tree

o   Once the development process and internal testing of the team for the deliverables tree is finished, each one of these will be validated with the client. For that, a *review and acceptance meeting will be convened.*

o   If the number of deliverables is considerable, the review can be separated into more than one meeting.

o   If the deliverable is rejected by the customer, the respective correction can be made in this same tree, if there is time in the tree, or add to the next tree, if there is no time available in it.

o   If the customer accepts all deliverables, the deliverables tree is considered closed and formally accepted.

o   If the tree is closed, it returns to point 2, to continue with the following deliverables.

o   If the decision has been made to define more requirements throughout the project, the necessary requirements must be redefined and prioritized, to start again in point 2.

Step 10: Lessons from the Deliverables Tree

o   After the deliverables tree is approved and accepted, *the team meets and analyzes the actions executed in each of the previous steps, to determine and what aspects are necessary to improve for the next tree.*

o   All lessons and improvement actions are added to *the Iteration N Lessons document.* This document will also maintain the history of each tree with its start and end dates and the total number of deliverables accepted and rejected. As shown in the following example:

N° Iteration	2
Start date	02.03.2020.
End date	15.04.2020
Number of Tree Deliverables	10
Accepted deliverables	7
Rejected Deliverables	3
Lessons learned.	

Lesson 1

Lesson 2

Lesson N

The *tree of deliverables* completed after executing the steps discussed above, is shown in the following figure.

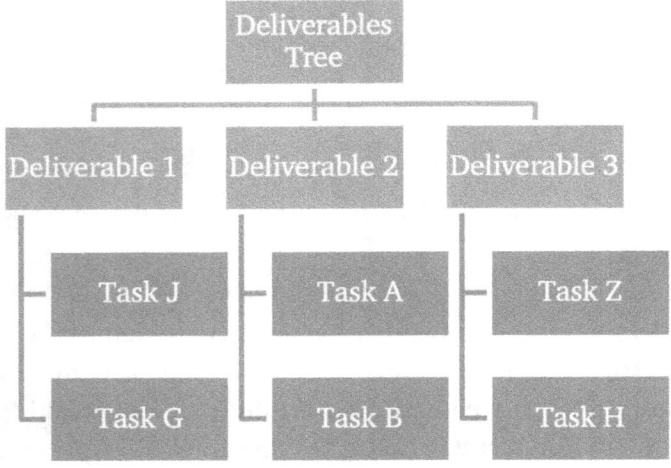

Figure 25 Deliverables Tree

Note

You should never have the same team tackling two deliverable trees at the same time, as you will lose focus on the deliverables.

The Product and Deliverables

Each time a *deliverable tree* is accepted by the customer, you will be increasing the perceived value of the product incrementally, generating a perception of greater value after each delivery.

Control your Deliverables.

A direct and simple way to control your deliverables is through a Kanban board, where you arrange the tasks according to what part of the development process, they are in.

This representation allows you to graphically see the status of tasks and quickly analyze tasks that are stopped, delayed, or not prioritized correctly in your development process.

Pending (12)	In progress (5)	Completed (2)	Accepted (3)	Cancelled (1)
Task 1	Task 3	Task 5	Task 7	Task 8
Task 2	Task 11			
Task 23	Task 10			
Task 4			Task 6	
Task 9	Task 15	Task 7	Task 21	
Task 16	Task 20			
Task 22				
Task 24				
Task17				
Task 14				
Task12				
Task13				

Figure 26 Task board

Note

The greater the number of deliverables and tasks that the team works on in parallel, the greater the complexity of the project to prioritize the tasks that need to be addressed to reduce turnaround time.

Prioritize and control the tasks of your Deliverable.

Here are some tips for prioritizing and controlling the tasks in your deliverable:

o   Prioritize the tasks needed to develop the deliverable.

o   Tackle one task at a time, finish it completely before starting the next one.

o   Do not work tasks in parallel (*multitasking*), it is less productive.

o   Keep the focus on the tasks that have been prioritized for the deliverable.

o   The team must be autonomous in determining how to assign tasks based on the profile and capacity of its members.

o   Monitor daily the progress of tasks with the team, using Kanban or another board for task control.

Control the value of your deliverables.

To control the value of each deliverable, it is important to consider the following points:

o   Control and prioritize the tasks of each deliverable.
o   Eliminate unnecessary expenses (*strategy 9*).
o   Monitor the impact of changes on deliverables.
o   Control the risks of the deliverable (*strategy 7*).

The variables that impact the value of the deliverable are shown graphically below.

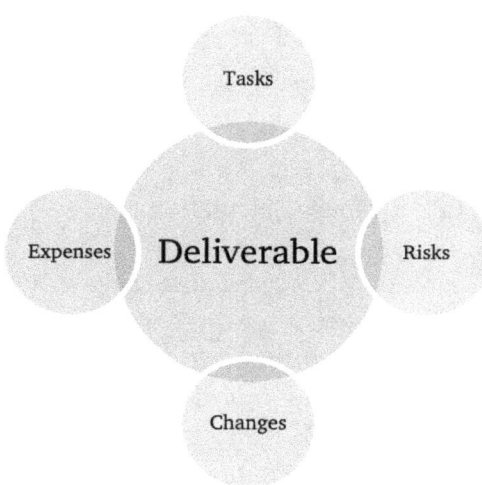

Figure 27 Variables that impact deliverables.

Each of these deliverable variables can be reviewed in follow-up meetings, both with the team and with the client.

Case Study

For this case study, we will define all the steps mentioned for the creation of a *deliverable tree*.

RET INC

For the creation of the *deliverables tree,* we will use as an example the requirement: *"Calculation of conversion effectiveness",* previously reviewed in the previous chapter.

Requirement Detail

1. Conversion Effectiveness Calculation

    1.1. The calculation of effectiveness seeks to solve the problem, of whether a prospect, after three months will become a client of the company. This currently leads to errors in revenue estimates for each quarter of the company.

    1.2. The entry of the information of the prospects is carried out by the four managers of client accounts, during the three months prior to the closure of the system for its calculation.

    1.3. The calculation of effectiveness is carried out by the head of the commercial area, taking the list of prospects entered in the GFT relational database of the commercial system, composed of:

    - o   The name of the package leaflet (N)
    - o   Time on the list since entry (T in months)
    - o   Level of interest (scale of 1 to 3, with 3 being the most interesting) (I)
    - o   Decision-making power (scale of 1 to 3, with 3 being the most powerful) (P)
    - o   Budget (B) (scale of 1 to 2, with 2 budget approved and 1 not yet approved) (P).

- o Importance of the product to the company (scale from 1 to 3, with 3 being the most important) (M).
- o Financial situation of the prospectus company (value to calculate Income/Debt) (D).

1.4. The head of the commercial area applies the following formula to calculate the conversion effectiveness value (VEF).

For a prospect N its value would be:

$$FEV= 0.5*(P* M* 0.5 I* D* (3/T)) +0.5* B$$

Calculation example:

$$FEV= 0.5*(P* M* 0.5 I* D* (3/T)) +0.5* B$$

N	T (months)	I	P	B	M	D	FEV	% FEV	Converted Client
Pros1	4	3	2	2	3	1	4,375	2,1%	No
Pros2	2	3	3	1	1	4	14	6,9%	No

The highest expected value of the calculation will be:

N	T (months)	I	P	B	M	D	FEV	% FEV	Converted Client
Pros1	1	3	3	3	3	10	205	100%	Yes

And the lowest expected value will be:

N	T (months)	I	P	B	M	D	FEV	% FEV	Converted Client
Pros3	12	1	1	1	1	0	0,5	0%	No

1.5. The VEF value is displayed every quarter, in a list that is distributed by mail to the Finance and Commercial department. The mail is issued by the head of the Commercial area.

1.6. It is expected that the system will take less than 5 minutes to calculate a total of 100 prospects, after the calculation is completed send an email with the calculated results of FEV to the Commercial and Financial departments. Additionally, the result must be loaded into the GFT relational database of the Commercial system. And finally, the calculated value must be available on the website designed for this new system and its mobile application.

1.7. The time required for the project from its inception to production will be 3 months.

1.8. The current FEV formula may be modified in order to achieve better adherence to the actual conversion data.

1.9. Historical data from 2013 to date FEV, both FEV estimates and actual conversions, will be available for unit and preliminary testing.

1.10.   To accept the deliverable, the value obtained from the VEF calculation will be compared at the end of the third month with the real customers who have converted. For this analysis, a percentage calculated on 85% of VEF as converted customer will be considered.

Step 1: Define the Iteration Cycle and Tree Validations

- o Iteration Cycle   : 6 weeks
- o Tree validation   : 3 days
- o Control meetings : Biweekly

Step 2: Define Deliverables (Results)

Three deliverables are defined:

Deliverable 1: Data Available for Calculation

- o Requirement:
  - ▪ Conversion effectiveness calculation.
  - ▪ Taking the points of the requirement: 1.1, 1.2 and 1.3.

- o Acceptance criteria

- Have the input data converted, without errors and ready to use in the creation of the new formula.

Deliverable 2: New EFV Calculation Formula

o Requirement:
- Conversion effectiveness calculation.
- Taking the points of the requirement: 1.4, 1.5 and 1.6.

o Acceptance Criteria:
- Validate the new EFV formula, reviewing the detail of the tests by which the formula was selected with respect to the others.

Deliverable 3: Results of the EFV calculation in Excel.

o Requirement:
- Conversion effectiveness calculation.
- Taking points from requirement 1.6 to 1.10

o Acceptance Criteria:
- Be able to have the results of the calculation in Excel, email, and the project database.

Step 3: Define a Deliverables tree.

Next, we sort and prioritize the tree's deliverables.

- o Deliverable 1: Data Available for calculation.
- o Deliverable 2: New EFV Calculation Formula.
- o Deliverable 3: EFV Calculation Results in Excel

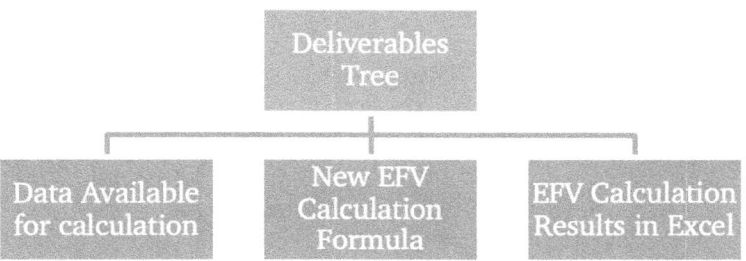

Figure 28 Deliverables Tree

Step 4: Create tasks for Deliverables.

In this step, we will create the tasks for each of the deliverables in the tree.

Deliverable 1: Data Available for Calculation

Tasks:

- o Task 1: Review access to source data.

- o Task 2: Convert and purge data from the source to the new *PROSPECT format*.

- o Task 3: Migrate data from GFT Relational Database of the Trading System to the *PROSPECT database*.

Deliverable 2: New EFV Calculation Formula

Tasks:

- o Task 4: Review components of the current EFV formula.

- o Task 5: Review historical data, calculation, and conversion of prospects, to determine correlation between the variables.

- o Task 6: Propose three new formulas.

- o Task 7: Compare EFV calculation data of the new formulas with the calculation and history of the current formula and select the formula with the best performance.

Deliverable 3: EFV Calculation Results in Excel and Email

Tasks:
- o Task 8: Display results in Excel.
- o Task 9: Send results by mail.
- o Task 10: Save results to PROSPECT database.

Next, we show how the tree of deliverables previously analyzed would look.

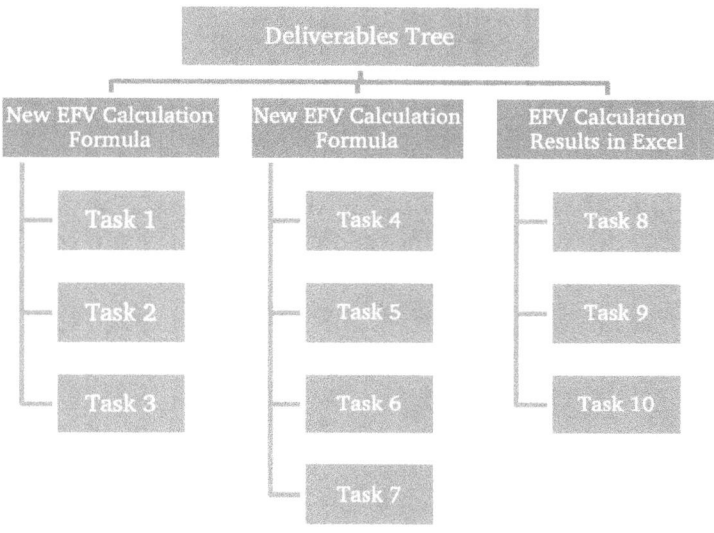

Figure 29 Deliverables tree with tasks

Step 5: Assign assignees to each task.

After all the tasks of the deliverables have been created, their managers must be assigned, and their estimates included.

Deliverable	Task	Responsible	Beginning	The end	Hours of effort
Deliverable 1	Task1	Jean Ro	12/03/2019	14/03/2019	16
Deliverable 1	Task 2	Jean Ro	15/03/2019	18/03/2019	24
Deliverable 1	Task 3	Hans Ils	18/03/2019	20/03/2019	24
Deliverable 2	Task 4	Robert Hlis	21/03/2019	22/03/2019	16
Deliverable 2	Task 5	Hortar Ju	23/03/2019	27/03/2019	40
Deliverable 2	Task 6	Robert Hlis	28/03/2019	05/04/2019	60
Deliverable 2	Task 7	Robert H	06/04/2019	13/04/2019	60
Deliverable 3	Task 8	External Provider	14/04/2019	18/04/2019	40
Deliverable 3	Task 9	External Provider	14/04/2019	18/04/2019	40
Deliverable 3	Task 10	External Provider	18/04/2019	20/04/2019	32

Total Deliverables Tree hours: 352

Step 6: Creating the Deliverables Tree

To begin the development creation of the *deliverables tree, the developments of* Tasks 1 and 2 *of* Deliverable 1 *begin*, which has the highest value prioritized by the client, as shown in the following figure:

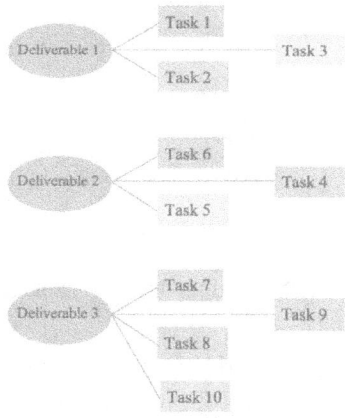

Figure 30 Creation of deliverables

Step 7: Execution and monitoring of the Deliverables Tree

During the execution and monitoring of the tree, the Kanban board of the tasks is reviewed together with the team, as shown in the following figure:

Pending (8)	In progress (3)	Completed (0)	Accepted (0)	Cancelled (0)
Task 1	Task 3			
Task 2	Task 11			
Task 23	Task 10			
Task 4				
Task 9				
Task 16				
Task 22				
Task 24				

Figure 31 Task statuses

In this step, you must validate each of the tasks that are presenting some inconvenience to advance and define actions to unlock them.

Step 8: Testing and Correcting the Deliverables Tree

The developed deliverables must pass the tests and corrections, for their subsequent release to the client. In our case study, they will be *deliverables 1 and 2*, as shown in the figure below.

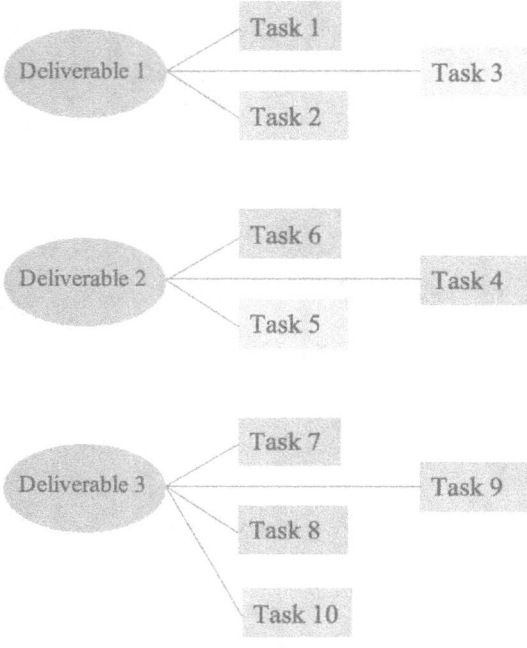

Figure 32 Deliverables in tests

Step 9: Validation and Closure of the Deliverables Tree

When the customer validates and accepts one or more deliverables, their tasks are passed to the accepted state on the Kanban board, as shown in the figure below.

Pending (0)	In progress (0)	Completed (0)	Accepted (11)	Cancelled (0)
			Task 1	
			Task 2	
			Task 23	
			Task 4	
			Task 9	
			Task 16	
			Task 22	
			Task 24	
			Task 3	
			Task 11	
			Task 10	

Figure 33 Accepted Tasks

In our case study, since all *deliverables* have been accepted by the client, all their tasks have moved to the accepted state, as shown in the figure above.

Step 10: Lessons from the Deliverables Tree

The lessons of the *deliverables tree* are:

- o Meetings
    - ▪ It is important to start the meetings at the agreed time, given the client's rigor regarding the issue of schedules.

- o Deliverables
    - ▪ An extensive test dataset is required for the validation and acceptance session of the deliverable with the customer.

- o Team
  - The team must improve communication and time estimation for integration tasks to create the deliverable.

STRATEGY 7: *MANAGE YOUR RISKS*

"If the project does not register risks, the project has a great risk."

The objective of this strategy is to systematize the way to record and manage risks during the development of your project.

Risks must be collected and declared throughout the entire project, from requirements gathering to product launch.

Each of the risks must be associated with a person in charge who carries out actions for its management.

If risks are not managed in a timely manner, they can generate an alternation in the project in terms of its time, cost, and quality variables.

What is a risk?

We will understand by risk, any condition of positive or negative uncertainty that may affect the development of your project.

This uncertainty condition could be controllable, uncontrollable, or unknown.

Discover the risks.

Risks are defined and analyzed every week, as defined in the field of Management and Communication.

Both in the case of *adaptive projects* and *predictive projects*, the start of *risk management* begins from the moment you define the requirements.

It is highly recommended to approach the risk discovery process from a broad to a detailed perspective. Begin by analyzing risks from the overall project view, then move to the requirements, followed by the deliverables, and finally the tasks. The following figure illustrates this recommended strategy.

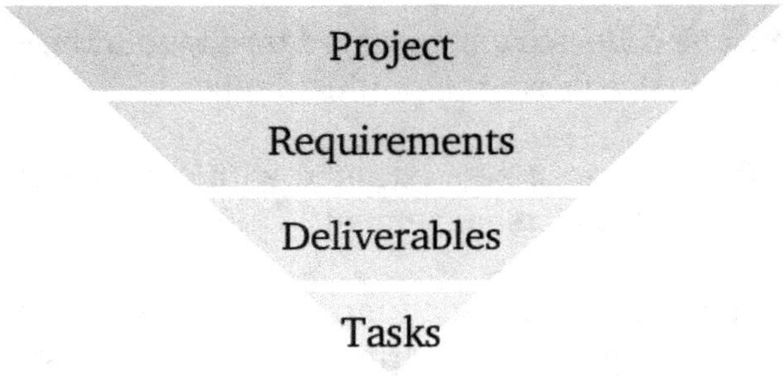

Project

Requirements

Deliverables

Tasks

Figure 34 Risk Analysis

During this discovery process, there may be risks that are repeated or similar, from each of the visions that are being analyzed, but then, during the detailed analysis of the risks, they can be merged or separated.

Tools to define risks.

To discover and define the *risks* of your project, you can use some of the following tools:

- o Brainstorming
- o Interviews and Questionnaires
- o Cause-and-effect analysis
- o Analysis of the Deliverables tree

Steps to Define the Risks of your project.

Here's a sequence of steps to help you define and record your project *risks*.

Step 1: Brainstorm

1.1. To start the process, brainstorm with the team focusing on the project as a whole and followed on each of the requirements raised.

1.2. Record all *risks* arising from this process.

Step 2: Review Deliverables Tree

1.3. For each of the project deliverables, it performs a review, to analyze possible associated *risks*.

1.4. For each of the tasks in the deliverables tree, perform an analysis of possible associated *risks*.

1.5. Record all *risks* arising from this process.

Step 3: Risk Register

When registering a risk, it must be analyzed from the point of view of the consequences and effects that this *risk* can generate to the project.

The *risk* register can have a structure such as the following:

o   Name of Risk
o   Risk Area
o   Responsible for Management
o   Impact on the project (Varies between 1-10)

  ▪   Low   1-3
  ▪   Medium   4-7
  ▪   High   8-10

o   Probability of Occurrence in the project (%)

  ▪   Casualty   1%-30%
  ▪   Medium   40% - 70%
  ▪   High   80% - 100%

o   Variable that impacts risk

  ▪   Time
  ▪   Cost
  ▪   Quality
  ▪   Product (Functionality, Plan, Scope)
  ▪   Team

Risks should be recorded based on their *Impact and* Occurrence *for the project, weighting them in terms of the variables of* Quality, *Time,* Product, Team *and* Costs.

Step 4: Risk Control Actions

Once the risks have been prioritized, it is necessary to take action on them. Below, we detail a set of actions that you can apply to control *risks*:

o   Mitigate

    ▪   The goal is to reduce the *risk* from its initial state. For example: if the risk is the lack of training of the team in a specific technology, training the team would be a way to mitigate the risk.

o   Transfer

    ▪   The goal is to transfer risk outside the scope of the project. We continue with the example about the risk of lack of knowledge in a specific technology, we could hire an external provider to work on that technology that we do not have internally and thus externalize the risk.

o   Avoid

    ▪   Analyze, if possible, to perform any action that allows you to prevent the risk from occurring. We continue with the example about the lack of knowledge in a technology, if we wanted to avoid it, we would have to change to a technology where the team has the necessary knowledge, thus avoiding the risk.

o Accept

- There will be risks where no action can be taken, other than accepting it and determining the impact and occurrence it will bring to the project, in terms of time, quality, product, Team and project costs.

Risk analysis

After recording each of the risks, it will be necessary to prioritize them, based on the valuation we have assigned.

These assessments should be reviewed on a weekly basis and may vary depending on the various situations and progress of the project.

Next, we will detail a record and analysis of *prioritized* risks.

List of Risks

1. Risk 1: Immature Technology
2. Risk 2: Change of client During the project

Risk Register

N	Area	Assigned	Probability (P)	Impact (I)	Impact Variable	Prioritized Value (P*I)
1	Technological	Helf Frag	85% High	8 High	Product Quality	6,8
2	Customer	Joel Gratf	10% Casualty	10 High	Product Time Cost	1

From the table above, we can infer that the first risk should be prioritized for attention due to its higher probability and greater impact on the project.

It is worth mentioning that the values $(P*I)$ may vary between the following values:

- The maximum value $(P*I)$ of a risk shall be 10.

- The minimum value $(P*I)$ of a risk shall be 0,01

Risk charts

By graphing the variables of *impact* and *probability of occurrence* of risks, we can achieve a clearer vision of the *risks* of your project.

Name	Probability	Impact	Prioritized Value P*I
Risk 1	10%	10	1
Risk 2	30%	8	2,4
Risk 3	40%	5	2
Risk 4	60%	9	5,4
Risk 5	80%	5	4
Risk 6	20%	6	1,2

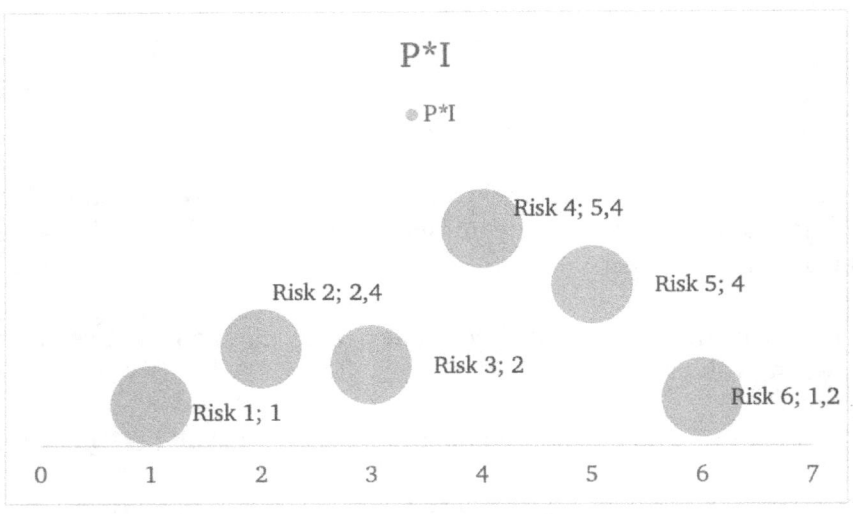

Figure 35 Risk Analysis by Impact and Probability of Occurrence

In the graph shown, risk 4 is the one with the highest priority to be addressed.

Case Study

RET INC

During the *PROSPECT project*, the following risks have been identified.

Next, we will review the entire process carried out for *risk management*:

Step 1: Brainstorm

At Project Level

The risks identified at the project level were as follows:

- The estimated time of 3 months for the delivery of the product may be too short. What can generate a partial delivery of the product in the required time.

- Part of the project team and the client are offshored, which can increase coordination and management times.

At the level of Requirements

When defining the *Conversion Effectiveness Calculation requirement*, the *risks* listed were:

- Quality of the input data from the Commercial Database.

- Effectiveness of the calculation formula in the required times.

Step 2: Review Deliverables Tree

At the level of Deliverables

The risks detected at the level of deliverables were the following:

Deliverable 1: Data Available for Calculation

- The lack of data from the Commercial database can affect the quality of the formulas.

Deliverable 2: New EFV Calculation Formula

- Too little time is detected to find a new calculation formula, this can cause reduced effectiveness.

At the task level

Task 1: Review data source

Task 2: Convert and cleanse data from the source into the new PROSPECT format.

- Data cleansing may require additional time, without customer support.

Task 3: Migrate data from GFT relational database of the Trading System to the *PROSPECT database*.

- A commercial database engine expert will be required for migration, as it is a COBOL engine.

Task 5: Review historical data, calculation, and conversion of prospects, to determine the correlation between the variables.

- Low quality and errors in historical data.

Task 7: Compare EFV calculation data of the new formulas with the calculation and history of the current formula and select the formula with the best performance.

- The analysis time of the formulas can be limited.

- Historical data may be missing that delays the process.

Step 3: Risk Registration and Prioritization

Next, we display the list and record of each of the *Risks* analyzed.

List of Risks

1. The estimated time of three months for the delivery of the product, may be too short, which can generate the partial delivery of the product in the required time.

2. Part of the project team and the client are offshored, which can increase coordination and management times.

3. Lack of input data from the Commercial Database can affect the quality of formulas.

4. The analysis of the calculation formula in the required times is reduced, this can cause a reduced effectiveness.

5. Cleaning input data may require additional time, without customer support.

6. A commercial database engine expert will be required for its migration since it is a COBOL engine.

7. Low quality and errors in historical data.

8. The time of analysis of the formulas can be limited to select an alternative.

9. Historical data may be missing that delays the validation process.

Registration and Risk Analysis

N°	Area	Assigned	Probability (P)	Impact (I)	Impact Variable	Prioritized Value (P*I)
1	Product	Greck Gast	85% High	10 High	Product Time	8,5
2	Team	Greck Gast	50% Stocking	5 Medium	Time Cost	2,5
3	Product	Helf Frag	90% High	4 Medium	Quality Time	3,6
4	Product	Joel Gratf	90% High	10 High	Quality Product	9,0
5	Product	Helf Frag	50% Stocking	5 Medium	Time Cost	2,5
6	Technolog ical	Joel Gratf	100% High	10 High	Cost Time	10
7	Product	Joel Gratf	50% Stocking	6 Medium	Time Product	3.0
8	Product	Joel Gratf	90% High	4 Medium	Quality Time	3,6
9	Product	Joel Gratf	50% Stocking	6 Medium	Time Product	3.0

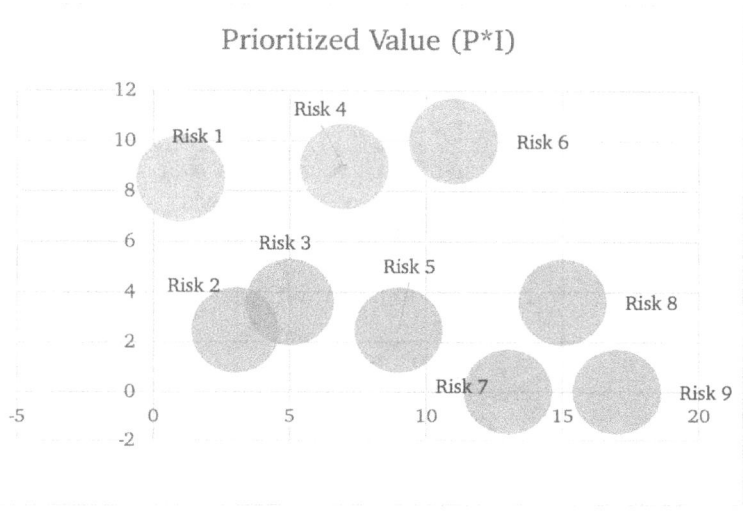

Figure 36 Prioritized Risks

From the graph above, it can be seen that risks 6, 4 and 1 should be addressed as a priority given their impact and occurrence in the project.

Step 4: Risk Control Actions

Taking the record and prioritization of *previously defined risks*, the following actions are defined:

N°	Action Type	Action Detail
1	Mitigate	Explain to the client the rationale for the risk "time too short for the entire product of the project". Propose as an alternative, make partial deliveries of the product, to monitor the impact of risk, after each deliverable.
2	Accept	The project team was conceived as delocalized, with a part of it in the same location of the client, so the risk will be present throughout the development of the project.

3	Transfer	The client must be informed that the lack of input data can affect the analysis of the current formulas and the quality of the new ones, so it is important to have the missing data or propose new ones to perform the analysis correctly.
4	Mitigate	Explain to the client, the impact of not having enough time to analyze all the formulas at the same time, so it is proposed to sequence the analysis and then test its effectiveness.
5	Transfer	Ask the client to perform the cleaning of their source data, to avoid interpretation problems and bad debugging by the project team.
6	Transfer	The migration of information from database will be carried out by an external provider that has expert COBOL personnel.
7	Transfer	The client is asked to perform the filtering of their historical data to avoid interpretation problems and bad debugging by the project team.
8	Mitigate	Explain to the client that the impact of not having enough time to analyze all formulas at the same time, so it is proposed to review only two formulas.
9	Avoid	The customer must be informed that, in the absence of historical data for a specific month, the comparison will only be made with existing data, to avoid delays in testing times.

STRATEGY 8: *QUALITY AND DELIVERABLES*

"If you produce them well from the beginning, you have advanced twice as far as possible."

The quality of the deliverables is an essential factor for the success of your project, since it is associated with the value that the product brings to the client. The value of the product is perceived every time you release a project deliverable.

Through this strategy, we seek to measure and control the quality of your product and Team, based on the deliverables of the project.

Each of the metrics that we will present in this chapter can be monitored at the time of releasing your deliverables, to make the necessary adjustments, in order to improve the quality of these.

The Impact of Errors on Quality

Errors *detected* in the product generate a significant impact on the quality and value perceived by the customer. Therefore, it is important to properly manage errors to improve product quality.

Active error management involves proactively managing errors throughout the development of the project, not waiting for the client to detect errors, but correcting them prior to the release of the product or deliverable.

If during the development process, you detect that it is necessary to stop the process, to correct an error, do it, correct the problem immediately. As we see in the figure below, the impact of correcting errors, in terms of time and costs, increases ostensibly in the final stages of the project.

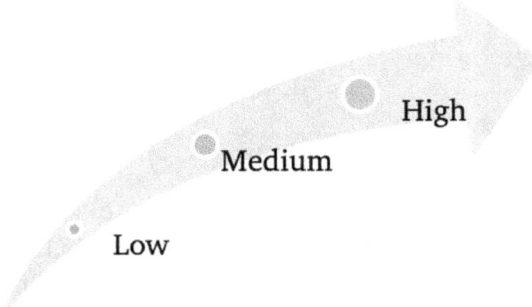

Figure 37 Impact of errors on the project

Team quality metrics

Depending on the time used, the rework performed, and the deliverables accepted, the performance of the Team can be measured considering the following metrics:

Percentage of Errors in Deliverables (PEE).

The ESP is calculated by dividing the number of errors detected by the client by the total errors detected in the deliverable, the latter corresponding to the errors detected by the client plus the Team, as shown below:

PEE = (Errors detected by the client /
Total errors detected (computer + client)) *100.

The values of the ESP that can vary in the following ranges:

- 0%   The customer perceived the deliverable without errors.
- 50%   The customer detected 50% of the errors.
- 100 % The client detected all errors.

The higher the percentage of PEE, the lower the perception of quality perceived by the client and, in turn, the lower the quality of the developments released by the team.

Percentage of Hours Invested in Deliverables. (PHE)

The PHE is measured by dividing the actual hours invested in the development of the accepted deliverable by the planned hours of this, as shown below:

PHE = (Actual hours accepted / Planned hours) *100

- <= 100%   sin gaps with the plan
- > 100%   with time lags

PHE values less than 100% indicate better Team performance, as well as improvements in the generation of deliverable estimates.

Reverse Rework Hour (HRI)

The HRI considers the difference between the actual hours invested and the hours initially planned for the deliverable, throughout its cycle, which includes from definition to approval.

HRI = Actual hours invested – Planned hours.

- $<=0$ without Rework
- >0 con reworks

If the HRI value is greater than zero, it indicates that a number of rework hours have been performed for the deliverable, if it is less than zero, the hours are within the schedule.

Proportion of Deliverables Rejections (PRE)

The PRE value is calculated by dividing the number of rejected deliverables by the number of accepted deliverables, as shown below:

PRE = Total deliverables rejections / Total deliverables accepted

- 0-1   It has no rejections for accepted deliverables.
- 1-10   It has rejections as accepted.
- >10   Too many rejections for each deliverable.

If we get a PRE value of 5, this indicates that for every 5 rejected deliverables, we will have 1 accepted deliverable. The lower the number, the greater the effectiveness of the Team and perceived quality of the product.

Proportion of Accepted Deliverables (PEA)

The PEA value is calculated by dividing the number of accepted deliverables by the total deliverables, the latter includes the most rejected accepted, as shown below:

PEA = Accepted deliverables / Total deliverables (accepted + rejected)

- 1   accepted without rejection.
- <1   accepted with rejections.

Note

A deliverable that has been rejected indicates that hours will have to be reinvested in the deliverable that does not meet the acceptance criteria or lacks to develop some functionality. This generates a rework on the part of the team.

Deliverable quality metrics

The quality of the deliverables will be given by the adherence of the deliverable to its acceptance criteria, functionality and expected benefits for it. Here are some metrics for the quality of deliverables:

Functionality

The value is determined by validating if the deliverable meets the approved functionality.

- 5   Meets with all functionalities.
- 3 - 4   Partially complies.
- 1 - 2   nor complies.

Acceptance Criteria

The value is calculated by determining if the deliverable meets the acceptance criterion.

- 5   Meet with the criterion, it is considered accepted.
- 3 - 4   partially compliant.
- 1 - 2   is rejected.

Expected Benefits

The value is calculated determined if the deliverable meets the expected benefits.

- 5   Meets expected benefits between 91-100% of the initial profit.

- 3 - 4   partially compliant between 40-90%.

- 1 - 2   non-compliant less than 30%.

The benefits obtained by the deliverables are contrasted with the initial benefit estimate of the project.

Case Study

RET INC

The project has defined three deliverables, with the next evolution during the project, starting with an initial estimate of hours of effort and duration dates.

Initial

Deliverable	Beginning	The end	Hours of effort
Deliverable 1	12/03/2019	20/03/2019	30
Deliverable 2	21/03/2019	28/03/2019	40
Deliverable 3	02/04/2019	20/04/2019	60

Week 4 of the Project

Deliverable	Hours Spent	Rejected	Approved	Internal errors	Errors When releasing
Deliverable 1	40	1	1	10	12
Deliverable 2	60	2	1	7	-
Deliverable 3	90	3	1	9	-

Final

Deliverable	Hours Spent	Rejected	Approved	Internal errors	Errors When releasing
Deliverable 1	40	1	1	11	13
Deliverable 2	80	5	1	17	8
Deliverable 3	120	7	1	19	5

Team Quality Metrics

Week 4 of the Project

Deliverable	PEE	PHE	HR
Deliverable 1	55%	133%	10
Deliverable 2	0%	150%	20
Deliverable 3	0%	150%	30

PRE	PEA
6	0,17

For every six (6) rejections one (1) deliverable accepted.

Final

Deliverable	PEE	PHE	HR
Deliverable 1	54%	133%	10
Deliverable 2	32%	200%	40
Deliverable 3	21%	200%	60

PRE	PEA
4,3	0,19

For every four (4.3) rejections one (1) deliverable accepted.

Deliverable Quality Metrics

Deliverable	Functionality	Acceptance criteria	Proceeds
Deliverable 1	4	5	3
Deliverable 2	5	5	2
Deliverable 3	4	5	4

From the analysis of the table above, it is inferred that in terms of functionality and acceptance criteria the deliverables meet expectations, but when capturing the benefits, they fail to satisfy the initial estimates of the client.

STRATEGY 9: CONTROL YOUR EXPENSES

"If you spend more than necessary, you will be losing a part of the benefit of your project."

The Expense Control strategy is designed to provide you with the tools necessary to identify primary sources of expenses and to monitor and manage expenditures throughout the project.

Sources of expenditure

Among the main sources of project expenses, we can find:

Hours

o   The hours the team spends on tasks and deliverables. These hours are distributed throughout the project, depending on the deliverables to be released.

External Services

o   Services from suppliers and others for the project.

Trips

- o Displacements of the project team, for example: to define requirements, present deliverables, meetings, trainings, and any other management related to the project.

Purchases

- o They include the purchases of utensils and Team necessary for the development and implementation of the project.

Project Expense Analysis

It is essential to conduct a project expense analysis at the outset of the project, when defining the areas, to obtain an initial estimate of project expenses. This should then be followed by a second, more detailed estimate, once the Deliverables Tree has been defined. In order to ensure accuracy, it is vital to consider all potential sources of expenses:

- o Hours
- o Purchases
- o External Services
- o Trips

Here's an example of an initial expense table for a project.

Initial analysis of project expenses.

Fountain	Amount USD	State
Hours	29.000	Initial
Purchases	15.000	Initial
External Services	15.000	Initial
Trips	10.000	Initial

Total expenses: 69,000 USD

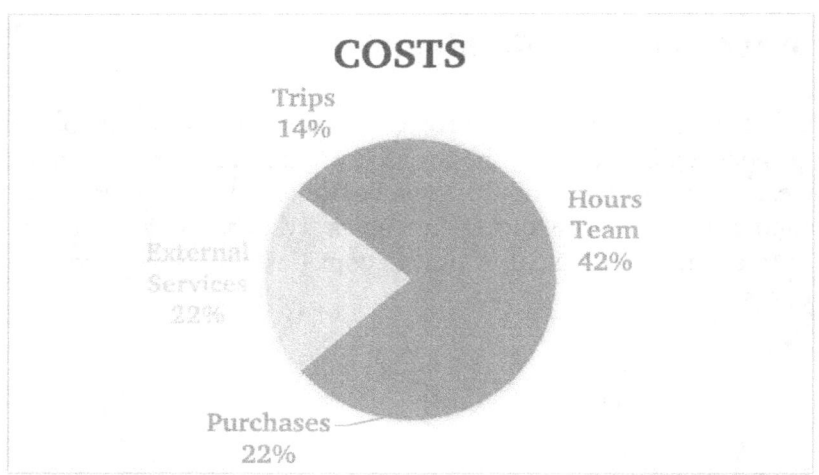

Figure 38 Project Expenditure Ratio

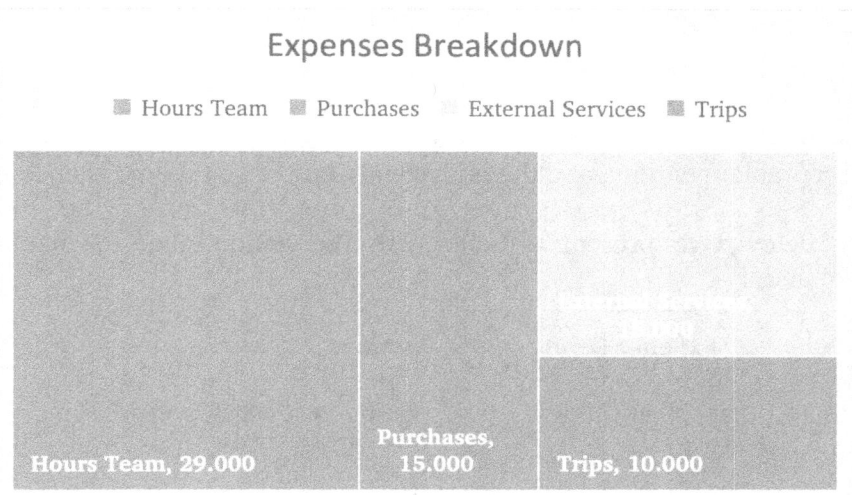

Figure 39 Breakdown of Expenses

Once you've defined your deliverables, you'll need to distribute your estimated expenses across each of them, based on their source, as shown below:

Deliverable	Fountain	Amount USD	State
Deliverable 1	Hours	10.000	Initial
Deliverable 1	Shopping	9.000	Initial
Deliverable 1	External Services	15.000	Initial
Deliverable 1	Displacement	7.000	Initial
Deliverable 2	Hours	9.000	Initial
Deliverable 2	Shopping	6.000	Initial
Deliverable 2	External Services	0	Initial
Deliverable 2	Displacement	3.000	Initial

This initial expense table will be your base estimate of analysis for the development of the project.

Define Action Plan for Expenses

Next, you'll need to define an action plan for various levels of acceptable spending, and the possible actions you'll take for each level.

Below, we present a table with the action plan to manage expenses:

Expense Levels	Actions
< 20 %	Review and adjust sources of increased expenses.
20% - 60%	Replan your initial spending plan, review profitability and adjust your increasing sources.
>60%	Stop the project and validate whether it is economically feasible to continue.

Cost Savings Plan

Analyzing each of the expenses, in which we have incurred, we can define actions that allow to reduce the expenses of the project over time.

Here are some topics to include in your *expense savings plan*.

Decrease Rework

o The objective is to reduce rework in each of the delivered deliverables developed, through actions such as the following:

- Improve requirements specification to avoid misunderstandings.

- Improve internal testing.

- Proactive error correction, prior to delivery to the customer.

- Group pending functionalities in the same deliverable or tree.

- Define deliverables that can be developed during the same iteration of a deliverables *tree*, reducing development times and subsequent integrations.

Travel Plan

The objective of the travel plan is to define the types and quantity of trips at the beginning of the project, thus limiting the expenses associated with the project, as shown in the following example:

Travel plan

- o 10 sessions to define requirements.   (week 1 and 2)
- o 10 sessions to present deliverables.   (week 6 and 7)
- o 2 weeks of training.   (week 8 and 9)
- o 1 week Displacement start-up.   (week 10 and 11)
- o 20 control meetings with the client.

Any additional displacement must be added as a change to the project.

External Services

The goal is to reduce initially unplanned spending by reducing the variability of external services.

One strategy is to price external services for the entire project or for each iteration, and the hours of deliverables that will need to be addressed by the external provider, thus setting the expenditure on external services.

Purchases

The recommendation is to price with suppliers at the beginning of the iteration or start of the project for all the volume needed, to get discounts on the volume price. With this we ensure the supply and reduce unnecessary expenses over prices.

Analysis and Control of Expenses

The *control of expenses* is reviewed biweekly in the project, in order to validate, if any of these items increase disproportionately to take corrective actions.

For this process, the detail of each of the expenses incurred must be listed and their origin analyzed, as shown below:

Expense Analysis

The *analysis of expenses* is done by deliverables and then by expense source, so we will have a grouped list of expenses as shown in the following table:

Deliverable	Item	Detail	Amount USD	Status	Date
Deliverable 1	Hours	8 hours Task 1	10.000	Executed	23/03/2019
Deliverable 1	Displacement	Meeting Definition of requirement 1.	1.000	Executed	24/03/2019
Deliverable 1	Hours	8 hours Task 3	10.000	Executed	26/03/2019
Deliverable 1	Hours	8 hours Task 2	10.000	Executed	29/03/2019

Analysis of Expenses by sources

To perform expense *analysis by source,* we will group the deliverable expenses by their source, as shown in the following *expense tracking table* for a deliverable:

Deliverable	Item	Executed USD	Initial USD	Balance USD	Executed %
Deliverable 1	Hours	7.000	10.000	-3.000	70%
Deliverable 1	Shopping	6.000	9.000	-3.000	67%
Deliverable 1	External Services	20.000	10.000	10.000	200%
Deliverable 1	Displacement	12.000	2.000	10.000	600%
Total		45.000	31.000	14.000	145%

In the *previous monitoring table,* more than estimated has been spent on *travel* and *external services* items. The expense detail table, specifically the displacement item, will need *to* be inspected to review the additional reason for the expense.

Next, we'll review a case study to explain the project expense management process.

Case Study

RET INC

The project has defined its initial spending plan, its spending action plan, and its savings plan, considering the three deliverables that you must release, as shown below:

Initial spending plan

The first cost estimate for the project is as follows:

Item	Amount USD	Status
Hours	40.000	Initial
Purchases	10.000	Initial
External Services	12.000	Initial
Trips	5.000	Initial

Total expenses: $ 67,000 USD

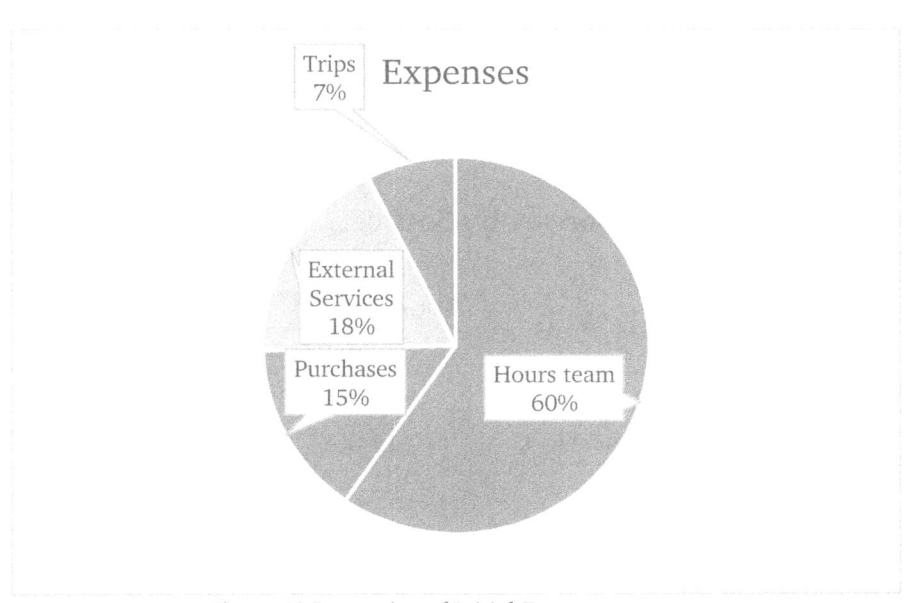

Figure 40 Proportion of Initial Expenses

153

After defining the deliverables of the project, the base estimate of expenses is as follows.

Detail of Expenses by Deliverable

Deliverable	Item	Amount USD	Status
Deliverable 1	Hours	10.000	Initial
Deliverable 1	Purchases	9.000	Initial
Deliverable 1	External Services	15.000	Initial
Deliverable 1	Trips	7.000	Initial
Deliverable 2	Hours	9.000	Initial
Deliverable 2	Purchases	6.000	Initial
Deliverable 2	External Services	0	Initial
Deliverable 2	Trips	3.000	Initial
Deliverable 3	Hours	11.000	Initial
Deliverable 3	Purchases	0	Initial
Deliverable 3	External Services	0	Initial
Deliverable 3	Trips	1.000	Initial

Summary of Expenses by Source

Item	Amount USD	Status
Hours	30.000	Initial
Purchases	15.000	Initial
External Services	15.000	Initial
Trips	11.000	Initial

Total Expenses: 71,000 USD

Expenses

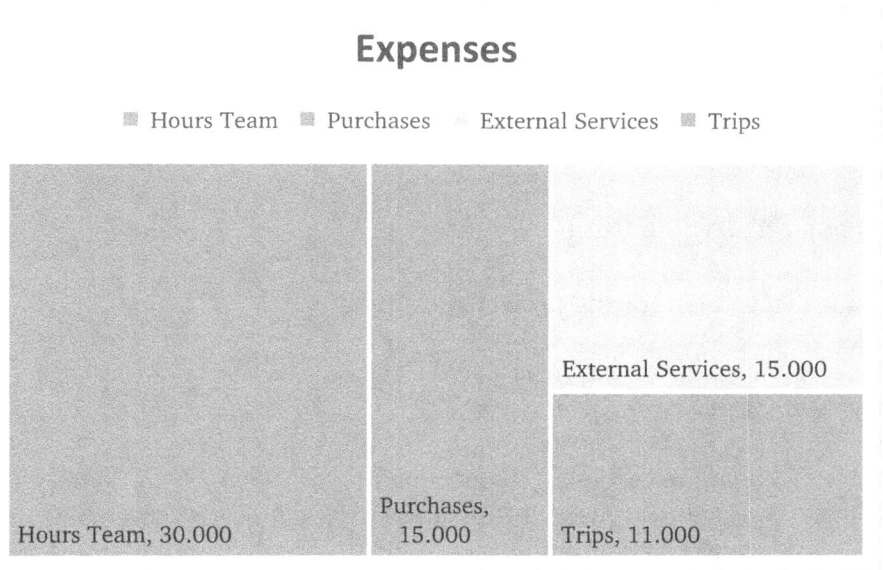

Figure 41 Initial Expense Detail

Expenditure Action Plan

The *expenditure action plan* includes the following actions:

Expenditure levels	Actions
< 10 %	Review and review sources of expenses that have increased and propose corrective actions.
10% - 60%	Review profitability and review causes of the increase in specific sources.
	Propose action plans for increasing sources, for example: freeze displacement on 50% increase in spending.

155

>60%	Stop the project and validate if it is economically profitable to continue with this level of spending.

Cost Savings Plan

The savings plan for the project considers:

Travel Plan

- o 7 sessions to define Requirements.
- o 6 sessions to present Deliverables (two for each deliverable)
- o 1 week of Training
- o 2 weeks Displacement for start-up
- o 20 control meetings with the client

External services

A value of 100 USD/ per hour is agreed with the supplier for the entire project. It is defined that the supplier will participate in the development of the tasks of *Deliverable 1*.

Purchases

All project purchases are detailed, to be executed the first week before the start of each deliverable.

List of purchases for the start of the project.

- o 2 computers
- o 2 computer keyboards
- o 1 printer ink
- o 6 whiteboard pencils

Analysis and Control of Expenses

Below is the breakdown of project expenses during week 4:

Deliverable	Item	Executed USD	Initial USD	USD Balance	Executed %
Deliverable 1	Hours	12.943	10.000	-2,943	129%
Deliverable 1	Purchases	8.000	9.000	1.000	89%
Deliverable 1	External Services	10.000	15.000	5.000	67%
Deliverable 1	Trips	4.000	7.000	3.000	57%
Deliverable 2	Hours	8.560	9.000	440	95%
Deliverable 2	Purchases	3.000	6.000	3.000	50%
Deliverable 2	External Services	-	-		
Deliverable 2	Trips	200	3.000	2.800	7%
Deliverable 3	Hours	6.000	11.000	5.000	55%
Deliverable 3	Purchases	-	-		
Deliverable 3	External Services	-	-		
Deliverable 3	Trips	-	1.000	1.000	0%
	Total	52.703	71.000	18.297	74%

From the detail of expenses of week 4, it is necessary to validate with the team, to which the increase in hours for *Deliverable 1* is due and validate if the deliverable was accepted by the client.

After the analysis of expenses, it is important to take actions so as not to continue in a spiral of increasing hours, which can make the project economically unviable.

Summary by source

Item	Executed	Initial	Executed %
Hours	27.503	30.000	92%
Purchases	11.000	15.000	73%
External Services	10.000	15.000	67%
Trips	4.200	11.000	38%

From this summary, it follows that given the progress of the project with only 4 weeks and the overall spending level in the *Hours source,* you will have to monitor the Hours expense source, for the remainder of the project and if necessary, replan this expense with your client.

The gap between executed and initial expenditures to date, separated by expense sources, is graphically shown below.

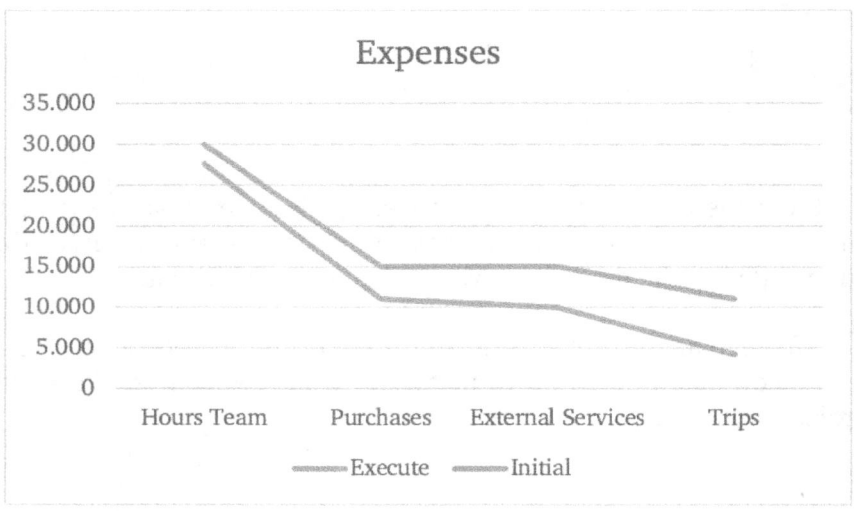

Figure 42 Executed vs initial expenses.

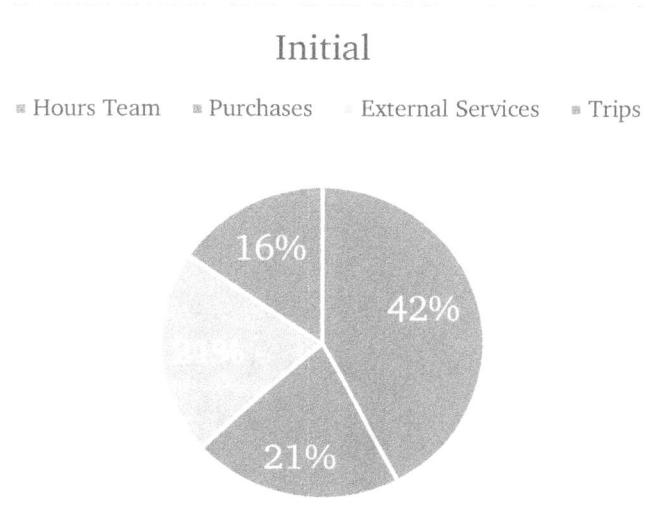

Figure 43 Breakdown of Initial Expenses

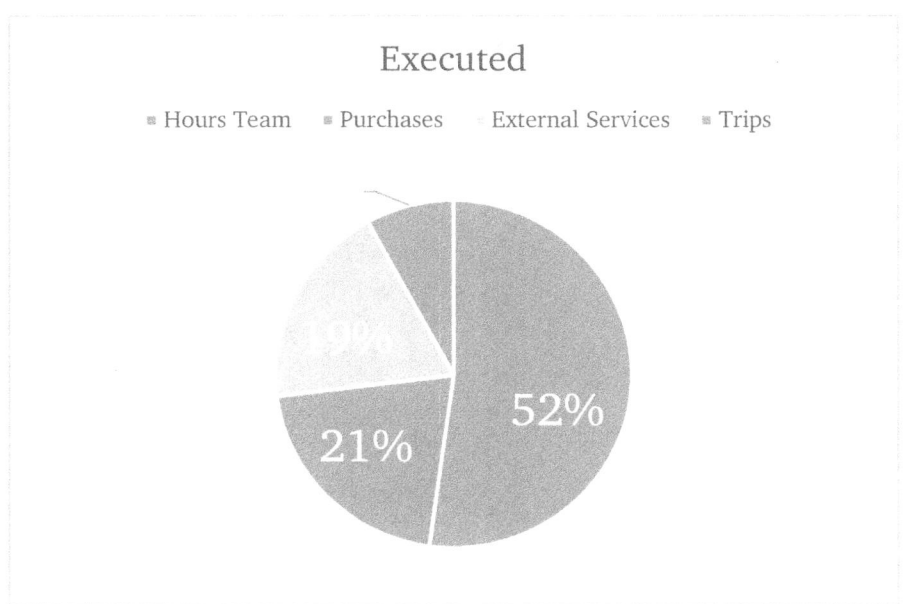

Figure 44 Breakdown of Executed Expenditures

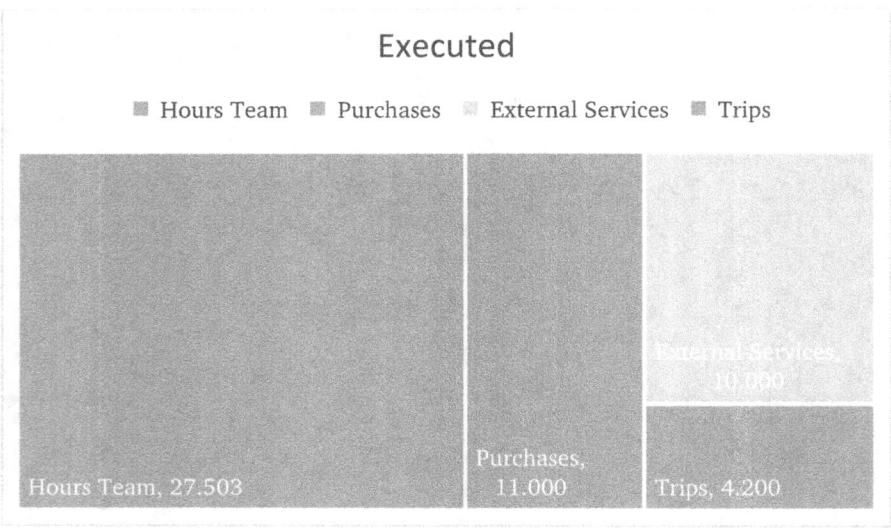

Figure 45 Breakdown Chart of Executed Expenditures

STRATEGY 10: *CONTINGENCY PLAN*

"Before climbing the mountain, prepare your return plan."

After reviewing the above strategies, it is critical to develop a contingency plan to quickly address issues that may arise. Experience suggests that at least one crisis is inevitable during the execution of your project.

If you have reached a state where you are facing a set of problems that arise unexpectedly during the project, it is important to have a contingency plan, which allows you to solve as many problems as possible in a very short time.

Next, we'll look at the main sources of problems in a project.

Problem Generators in your Project

There are several factors that can generate a crisis in your project, among which are:

- o Lack of time.
- o Unrealistic delivery plans (underestimates).
- o Uncontrolled requirements change.
- o Lack of resources.
- o Too much uncertainty in the team about the tasks to be performed.
- o Lack of agreements with the client.
- o Unmotivated teams with no focus on the important issues.
- o Unmotivated customers and without clarity of the required product.
- o Low communication and collaboration between team members and client.
- o Low quality of deliverables.

Make your contingency plan.

To develop your contingency plan, we recommend taking the following steps:

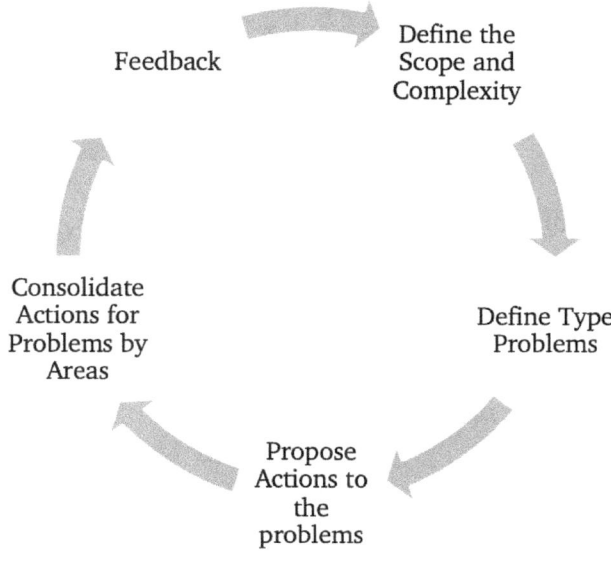

Figure 46 Contingency Plan

Step 1: Define the Scope and Complexity of your project.

It is important that, as a first step, you define your scope and complexity of the project, since this previous analysis will give you an initial knowledge of the issues, which may be relevant when defining and preparing your contingency plan.

Step 2: Define Type Problems

To define the "type" problems of the project, we can use two strategies that can be useful:

- o Take each of the areas of the project and using brainstorming, list the possible problems that are visualized in the project.

- o Another alternative is to start the analysis, based on the risks you have registered and place yourself in the scenario, in which each of them becomes a reality, as a latent problem for the project.

Step 3: Propose Actions to the problems.

For each of the scopes, it selects the registered problems and defines the actions that can be performed, given the scenario described in the problem.

Conducting an initial brainstorming session with the team is crucial to generate potential actions. Subsequently, refine and select the specific actions that can be incorporated into the plan.

Step 4: Consolidate Actions for Problems by Areas

The objective is to order all the problems and actions by areas, so that together they can form the contingency plan of the project.

Step 5: Plan Feedback

Each time you begin or complete the development of a deliverable, it is important to review during the lessons learned meeting whether there is a need to add new actions or address any issues in the plan.

When you need to use any of the actions, it is important that you determine if you need to adjust or add any additional actions.

This feedback process is dynamic, as issues may arise throughout the project. It is essential to have actions in place to address these problems as they occur.

Contingency Plan

Below, we show you an example of a *Contingency Plan*, starting by listing the problems that may arise in each of the areas and the possible actions to be taken.

Client Scope

Problem	Action
The client is not clear about the structure of profiles and roles to address the project.	Prepare a proposal of profiles and roles necessary to tackle the project.
The client does not have the budget to address the entire project scope.	Propose: -A smaller scope, adjusted to the highest priority to meet your needs. -A project of the adaptive type, where you release deliverables depending on the time and money you have.

Team Scope

Problem	Action
One member leaves the team	Always distribute the tasks of the project in different members, communicate the new assignments to the rest of the team, inform the client of the situation, if the impact cannot be limited only to the team.
A team member is not clear about their tasks for the week.	In the internal coordination meeting, define with each team member the assigned tasks and the objective of the week.

Management and Communication Field

Problem	Action
The meetings extend more than two hours, without a clear objective on the topic being addressed. They have become a waste of time.	In the citation of the meeting send the detail of the topics to be discussed and the objective sought, this will limit the topics.
	Take control and leadership of the meetings based on progress on each of the planned topics. If one of them cannot be resolved, it generates a subsequent meeting to discuss that particular issue.

In meetings with remote teams or clients, the time difference causes errors in the meeting start time. For example, one team sees the meeting at 12:00 noon on the calendar and the other team at 1:00 p.m.	When you send a meeting appointment for a remote client or team, send it with the recipient's time zone, not the source zone, and add the time in both regions and countries in the title of the email.
The topics agreed at the last meeting have not been followed.	After each meeting, send a minute or mail of the assignments and agreements taken. Review them at the beginning of the next meeting to validate progress of the agreements.
Project risks appear spontaneously, and no one is managing them.	Review your risk strategy, if you do not find the answer, generate meetings of 30 minutes per week to detect risks, through brainstorming, write down and analyze with the team, the risks that each member visualizes in their work area of the project.

After writing down the risks, categorize them according to their impact and occurrence in the project, followed by assigning them to each of the team members, who will be responsible for making the necessary management to control it. |

Product Scope

Problem	Action
I am not in control of the deliverables, their status, allocation, and completion time.	Review your product scope and Communication, regarding deliverable control meetings. If you don't find the answer, define a 30-minute meeting at the beginning of the week with the team to review the tasks and assignments of the week for each deliverable.
The released deliverables have been rejected several times, for not meeting the acceptance criteria and lack of functionality	Meet with the team and review the causes, why the acceptance criteria cannot be met and the missing functionalities. To find the cause of the problems, check the following: o Validate if it is a technical development issue. o Lack of definition in tasks. o Lack of definition of the requirement. o Reinforce internal testing to proactively correct errors detected by the customer.

IMPLEMENTING THE STRATEGIES

"Apply the knowledge acquired, do not waste your time."

To conclude our presentation and facilitate the implementation of the strategies, we offer a summary of the Ten Strategies, along with accompanying templates for you to complete and apply to your project.

Strategy 1: Define Your Scopes and Complexity

Client Scope			
Name			
Role / Structure			
Actions and Decisions			
Location			

Team Scope			
Name			
Role / Structure			
Actions and Decisions			
Location			

Management and Communication Field			
Type of Meeting			
Detail			
Periodicity			
Participants			

Product Scope			
Product Management			
Detail			
Periodicity			
Responsible			

COMPLEXITY OF THE FIELDS			
Ambit			
Analysis			
Complexity (high, medium, low)			

Strategy 2: Create autonomous teams.

Define Your Autonomous Team		
Step 1: Define profiles		
Step 2: Set up your computer		
Step 3: Consolidate your team in a single location		
Step 4: Empower your team		
Step 5: Team and customer interaction		
Step 6: Inform your customer		

Strategy 3: Manage your customer.

List of Procedures		
Kick-off meeting		
Defining Requirements or Changes		
Control and monitoring of deliverables		
Review and acceptance of deliverables		
Customer Training		
Project Closure and Contract		

Strategy 4: Communicate 360

Communicate 360:	
Client Scope	
Requirements and Changes Sessions	
Control and Progress Meetings	
Validation meetings of deliverables and functionalities	
Team Scope	
Work Meetings	
Internal progress	
Product Scope	
Meetings to define Deliverables and Tasks	

Strategy 5: Effective Requirements

Strategy Definition Requirements:	
Step 1: Session Schedule	
Step 2: Session Participants	
Step 3: Define Questionnaires for requirements	
Step 4: Define the Core Problem	
Step 5: Requirements Detail	
Step 6: Requirements Document	

Questionnaires Requirements	
What?	
Who?	
How?	
Where?	
How much?	
When?	
Which?	

Strategy 6: Look for Results.

DELIVERABLES MANAGEMENT	
Step 1: Define the Iteration Cycle and Tree Validations	
Step 2: Define Deliverables (Results)	
Step 3: Define a Deliverables Tree	
Step 4: Create tasks for Deliverables	
Step 5: Assign task assignees	
Step 6: Creating the Deliverables Tree	
Step 7: Execution and monitoring of the Deliverables Tree	
Step 8: Validation and Closure of the Deliverables Tree	
Step 9: Lessons from the Deliverables Tree	

Strategy 7: Manage your risks.

RISK MANAGEMENT	
Step 1: Brainstorm	
Step 2: Review Deliverables Tree	
Step 3: Risk Register	
Step 4: Actions and risk control	
Step 1: Brainstorm	
Step 2: Review Deliverables Tree	
Step 3: Risk Register	
Step 4: Risk Control Actions	Mitigate Transfer Avoid Accept

RISK REGISTER TABLE		
Name of Risk		
Risk Area		
Responsible for Management (assigned)		
Impact on the project (Varies between 1-10)	Low Medium High	1-3 4-7 8-10
Probability of Occurrence in the project (%)	Casualty Stocking High	1%-30% 40% - 70% 80% - 100%
Variable that impacts risk	Time Cost Quality Product (Functionality, Plan, Scope) Team	
Prioritized Value (P*I)		

Strategy 8: Quality and deliverables

TEAM QUALITY METRICS	
PEE = Percentage of deliverable errors detected by the Customer / total errors detected (team + Customer)	5 Meets all functionality. 3 - 4 partially complies 1 - 2 does not comply
PHE = Percentage of actual hours spent / planned hours of deliverables.	<= 100% No lags from the plan > 100% has hour lags.
HR= Hour of rework reversed (measured from the definition, creation, release, rejection, and approval cycle). HR = Actual hours spent - planned hours of deliverables	0 no rework >0 with reworks
PRE = Ratio of rejected deliverables / accepted deliverables	0-1 It has no rejections for accepted. 1-10 Has rejections for accepted. >10 Too many rejections for each deliverable

PEA = Proportion of deliverables accepted / Total deliverables (accepted + rejected).	>=1 accepted without rejection <1 accepted with rejections.

DELIVERABLE QUALITY METRICS	
Functionality	5 Complies with all functionalities. 3 - 4 partially compliant 1 - 2 non-compliant
Acceptance Criteria	5 Meets all criteria. 3 - 4 partially compliant 1 - 2 non-compliant
Expected Benefits	5 Meets expected benefits between 91-100% of the initial profit. 3 - 4 partially complies between 40-90% 1 - 2 does not comply with less than 30%

Strategy 9: Control your spending.

ANALYSIS OF SOURCES OF EXPENSES					
Item	Amount	Initial	Executed 4, 8 16 weeks	Final Execution	Executed %
Hello-Purchases					
External Services					
Trips					

DETAIL OF EXPENSES	
Deliverable	
Item	
Detail	
Executed USD	
State	
Date	

EXPENSE ANALYSIS	
Deliverable	
Item	
Executed USD	
Initial USD	
USD Balance	
Executed %	

Strategy 10: Contingency Plan

DEVELOP THE PLAN	
Step 1: Define the Scope and Complexity of your project	
Step 2: Define Type Problems	
Step 3: Propose Actions to the problems	
Step 4: Consolidate Actions and Problems by Areas	
Step 5: Plan Feedback	

VOCABULARY

Project Scope: It consists of all the functionality detailed in the requirements that will be delivered as a product during the development of the project.

Change: Consists of any new product specification, not defined during the definition of the requirement. This new specification must be reflected in a *Requirement Change Document*, which must be approved by the client for subsequent execution.

Use cases: Detail of the use of the system or a part of it (a specific functionality), including the specification of all actors involved in the process.

Interested parties. They are all those people who directly or indirectly influence the project. For our context, we will separate these into two groups: Team and Customers.

Clients: They are all those people or entities to which the project affects them directly or indirectly. Among these can be found:

- o Sponsor
- o Users
- o Lead User
- o Key User
- o Project Office (PMO)
- o External entities
- o Internal and external regulatory bodies

Delocalized Client: This concept indicates that a part of the client's Team is in another geographical location (region or country) or physical location (different floor or building).

Duration: This corresponds to the deadline in terms of dates that a task or deliverable is performed. (hours)

Effort: This corresponds to the work (hours) needed to perform a task, in terms of time.

Team: It corresponds to all the people who must perform some task to produce some result in the project, such as:

- o   The main team,
- o   Third-party developers,
- o   Managers
- o   All those who work or will work for the project.

Delocalized Team: This concept indicates that the team is in another geographical location (region or country) or physical (different floor or building) with respect to the client.

Deliverable: Functional portion of the product to be developed, which corresponds to one or more business requirements, must be reviewed, and approved by the client before development.

- o   A deliverable must provide specific, measurable value to the customer.

- o   A deliverable is made up of a set of tasks that must be completed.

- o   Each deliverable has an acceptance criterion.

    o   Each deliverable must consider the quality criteria of the requirements it contains.

Strategy: We will understand by strategy the set of actions that seek to achieve a specific objective. For our context, the objective sought is the success of the project.

Kanban: This system consists of a whiteboard to control the sequences of the task states of the product deliverables. Therefore, it is a graphical way to see the status and control of tasks.

Project: A project comprises a series of temporary activities aimed at producing a unique product, service, or result. It is characterized by a defined start and end, a specific scope, and the allocation of both physical and human resources to complete the tasks and activities within the established timeframe.

Requirement: They are product specifications that address the business's needs and are detailed in the *Requirements Specification Document*.

Risk: It is any event that can occur in a certain period of time, which can directly or indirectly affect the correct execution of the project.

Task: It is a portion of time allocated to a specific activity of the project.

Development cycle or iteration: This consists of a certain window of time it takes to develop one or more deliverables of the project.

FIGURES

ABOUT THE AUTHOR

The author is a Computer Engineer and MBA with numerous specializations and certifications in Project Management.

He has extensive professional experience in Technology Project Management and Business Consulting, working with both startups and growing software companies.

He has led a variety of projects in different countries, managing local, remote, and offshore teams across various economic sectors. Additionally, he is the author of books on sales and technology projects.